Humanize Innovation

Human First. AI Forward.

Greg Brisco

Humanize Innovation
Human First. AI Forward.
Published by Humanize Innovation
www.humanizeinnovation.com
©2026 by Greg Brisco
ISBN #979-8-9988491-4-5
Library of Congress Number # 2026902737

Cover design by Rebecacovers on Fiverr

Published and printed in the United States of America.

Endorsements

"I love artificial intelligence, not because it replaces people, but because it has the power to reveal what people are capable of when barriers are removed. At its best, technology is the great equalizer of our time. It democratizes access to knowledge, opportunity, creativity, and voice in ways no previous era could imagine.

What Greg Brisco captures so powerfully in this book is the truth that AI is not the danger, disconnection is. When technology is guided by human-centered leadership, it becomes a force for inclusion, innovation, and shared progress. This book doesn't ask us to slow down the future. It challenges us to rise to it with intention, wisdom, and humanity.

Humanize Innovation is a timely and necessary reminder that the future doesn't belong to machines, it belongs to people who know how to use them responsibly, courageously, and for the greater good."

— Coach Michael Taylor

Human Potential Architect/Author

"Greg is an expert in AI, but even better, he knows how AI and the human dynamics of innovation need to be harnessed together in order to achieve the advances that are available to us. Integrating human systems with advanced technology is not simple, but it is what will make this new era of innovation adaptable and ethical."

Bob Rosenfeld, Founder and CEO, Idea Connection Systems, Inc., and creator of the Innovation Strengths Preference Indicator? (ISPI™) &

Andrew Harrison, Innovation Ambassador, Idea Connection Systems, Inc., and Master Trainer, Innovation Strengths Preference Indicator? (ISPI™)

"Greg Brisco is a true leader whose insights are grounded in a profoundly human and genuine philosophy. At a moment when the stakes for AI design and deployment could not be higher, this book is an essential, urgent blueprint. It doesn't just call for ethical AI; it shows us how to commit to ethical leadership and build a representative future from the foundation up."

Meme Styles, Founder & President of Measure, Chairwoman of HBCI AI Conference & Training

In "Humanize Innovation," Greg Brisco brings uncommon clarity and care to a truth this moment demands we confront: AI is not neutral, and leadership is the deciding variable. He thoughtfully articulates the Human Delta, the gap between ambition and human readiness, and invites leaders to design systems where accountability, values, and humanity are treated as essential infrastructure, not an afterthought.

Lori Holt, Founder & President of Novelle & Associates, The Thought Authority, and Co-Founder of Humanize Innovation

"In the era of AI, our existential challenge as a human race is whether this transformative technology will be leveraged to advance society or destroy it. If you care about ethical AI, this is a must read."

Dr. Tom C. Hogan, SHRM-SCP, GPHR, SPHR Professor of Practice in Human Resource Management at Penn State University, School of Labor and Employment Relations

Table of Contents

Acknowledgements: Before We Build, We Honor..................................... v

Introduction xi

What This Book Will Change for You xvi

Prologue: Human First, Leadership in Practice xviii

Chapter 1: The Velocity Problem, When Tech Outpaces Leadership 1

Chapter 2: The Human Delta™, Closing the Execution Gap........................ 7

Chapter 3: From Optimization to Humanization, Redefining the Edge......13

Chapter 4: Organizations Don't Innovate. People Do. 21

Chapter 5: Innovation Strengths Preference Indicator® (ISPI™),
Mapping the Human Elements of Innovation..............................27

Chapter 6: Engineering High-Velocity Teams, An Everyone,
Everywhere Imperative..55

Chapter 7: Innovation - Engineering a People-Driven
Innovation Engine..89

Chapter 8: Leadership - Re-Architecting Leadership for the AI Era...........97

Chapter 9: Community, Building a People-First Ecosystem 105

Chapter 10: Education, Reimagining Education in the World of AI 111

Chapter 11: Raytheon Innovation Challenge, Engineering Teams
for 10× Output...117

Chapter 12: VC Portfolio Operating System, Human-Centric
Investment..125

Chapter 13: Shared Responsibility, Aligning Stakeholders for
          AI Success 133

Chapter 14: Responsible and Ethical AI 141

Featured Case Studies & Partner Organizations 158

Books & Research Reports 159

About Greg Brisco 161

Acknowledgements

✦

Before We Build, We Honor.

To My Lord and Savior, Jesus Christ: Above all, I acknowledge my Lord and Savior, Jesus Christ, who is Lord of my life and the true author of this work. This book is not the product of my intellect, ambition, or discipline alone. It exists because of what You have placed inside of me and the way You guide me daily, often when I don't fully understand the path ahead. I would never pretend to take credit for what You have already gone before me to make possible. You open doors I could not open, sustain me when I am depleted, and refine me when I am resistant. Thank You for choosing me as a vessel, for trusting me with responsibility, and for allowing this work to be an expression of service to Your people. Everything I build, I place back at Your feet, because You are the creator and perfecter of humanity.

This book is dedicated to the leaders who understand that progress without humanity is not progress at all. To those willing to slow down and ask better questions. To those brave enough to lead when the answers are not obvious. To those who believe the future of artificial intelligence must be shaped by values, not just velocity.

Before we dive into the content, I want to acknowledge the people and partnerships that made this vision possible. My journey has been shaped by a community of mentors, collaborators, and friends who exemplify what it means to put humans first even as we forge ahead with technology. To them, I owe immense gratitude:

To the entire Idea Connection Systems (ICS) and Innovating Edge teams: Thank you for welcoming me into your fold and believing in this mission. I am especially grateful to Bob Rosenfeld, founder of the Innovation Strengths Preference Indicator® (ISPI™), whose pioneering vision and brilliance unlocked the diagnostic framework powering much of what you'll read here. Jonathan Rosenfeld, Andrew Harrison, Kenn Burke, and Dr. Jerry Fisher, thank you for entrusting me with your life's work and taking a chance on a bold idea. Being part of your story has meant the world to me.

To the Society for Collegiate Leadership & Advancement (SCLA) team: Your partnership is one of the greatest honors of my professional life. From our first conversation, you understood the urgency and power of humanizing leadership while serving as a career incubator for all learners. You not only saw the value in these ideas, you embodied it. Thank you for your trust and commitment to preparing the next generation of innovative leaders.

To Stedman Graham, Steve Jones, and the Identity Leadership team: Thank you for allowing me to carry forward a philosophy the world desperately needs. Identity is the starting point of all innovation. It's how we begin to heal what's fractured. Your work is sacred. I'm honored to play a small part in expanding its reach.

To Scott MacGregor and the team at Something New, and all the Outliers in the Outlier Project: You gave me the language and ethos that define this book: People Over Everything. That mantra is no longer just a business philosophy; it's the cornerstone of my life. Thank you for the friendship, the fierce honesty, and the example you set.

To Andrew Martin, Isaac Truong, and the entire Six-Figure Dinners and Allston Yale teams: Thank you for being the spark that pushed me to embrace entrepreneurship fully. I was only dabbling before, but your vision, dedication, and spirit inspired me to lean in with conviction. I'm especially grateful for that unforgettable retreat, for challenging me to be vulnerable and creating an environment that transcended business. You unlocked emotional and spiritual growth in me that I didn't even know I needed. From the bottom of my heart, thank you.

To the teams at the National Society for Leadership & Success (NSLS) and CNEXT: Thank you for giving me the opportunity to pivot into the world of leadership development, from collegiate leaders all the way up to Fortune 50 C-suites. What we built together laid a strong foundation for the values in this book. Collaborating with visionaries like Jack Stahl, Rick Smith, Cheryl Stokes, Chereé Huey, Neil Khaund, Chuck Knippen, Shahla Akbari, Amy Westby, Sean Anderson, Kristen Cooper, Iman Hill, and so many others gave me rich examples of excellence and humanity that infuse these pages. I cannot thank each of you enough for believing in me and giving me space to grow.

To Van Jones, Tadzio Smith, Gregory A.J. Ware, and the entire Rapport team: Thank you for your partnership and for sharing a vision of using AI to enhance the human experience at its core. The synergies we found from day one, and our shared commitment to human-first innovation, are exactly what this book is about. I'm honored to work alongside you in proving that technology can deepen human connection when guided by the right values.

To Joe Apfelbaum and the evyAI team: Joe, you represent what happens when generosity meets ingenuity. You didn't build evyAI to replace human voices; you built it to amplify them. Watching you demonstrate how AI can preserve a person's unique tone and perspective while removing engagement friction inspired me to think differently about scaling authenticity. You proved that in an algorithm-driven world, sincerity can still scale. Thank you for your friendship, your example, and for reminding me that service and heart are true differentiators.

To Lyndsay Dowd and the Heartbeat for Hire team: Lyndsay, you remind executives that culture is not a line item, it's the lifeblood of performance. You've redefined hiring by giving candidates not just access, but dignity and opportunity. Your work shows how technology and empathy can merge to create careers and cultures worth believing in. Thank you for proving that empathy doesn't slow innovation down, empathy drives innovation.

To Lori Holt, William "Classic" Jobs, Jacqui Rodrigues, and the entire Build With Lori/Thought Authority teams: You came into my life exactly when I needed you most, as thought partners, advisors, strategists, part-time therapists, and full-time friends. Your clarity and courage, and the community you've built around you, have made me a better builder, a better leader, and a better man. I can never repay you, but I will spend the rest of my career proving that your investment and belief in me were worth it.

To Telea Stafford and the Phenixx Marketing team: Telea, you and your team gave this work its name, and in doing so, helped give it its spine. Human First. AI Forward. did not emerge from branding exercises or clever wordplay; it came from your willingness to challenge me when it mattered most. You were strong enough to disagree, disciplined enough to push deeper, and grounded enough to protect the integrity of the vision while it was still being formed. That kind of leadership is rare, and it changed the trajectory of this project for the better. Thank you for holding the line on meaning, not just momentum. I have the utmost respect for you as a leader, strategist, collaborator, and as a fellow Texas Tech Red Raider, I couldn't be more honored to build alongside you. Your friendship is invaluable to me, and your imprint on this work will endure.

To the board and members of the Austin AI Alliance, and especially my fellow panelists Meme Styles (CEO of MEASURE) and Zen Van Loan: Thank you for embracing me and challenging me in our meetings and collaborations. I'm inspired by the work you do and grateful for the way you welcomed me (even joking that I'm "so Houston"). In Texas terms, we might be from different cities, but we are neighbors, and as far as I'm concerned, we're family. Your commitment to data equity and community-driven innovation is a guiding light.

To Daphne "Money" McDole, CEO of the African American Leadership Institute and CEO of Six Square: Thank you for your fearless leadership and unwavering pursuit of equality. You give a voice to those who too often go unheard. Your commitment to justice isn't about rhetoric, it's grounded in action and accountability to the community. You've taught me that this work isn't about recognition; it's about showing up, doing the work, and staying accountable. I am deeply grateful for the example you set and honored to work alongside you.

To Dr. Melva Wallace, President of Huston-Tillotson University, Chris Hyams, Board Member, and the entire Huston-Tillotson community: Thank you for your courageous leadership and for opening your campus and classrooms to a vision that extends far beyond any single conference or moment. Dr. Wallace, your commitment to positioning Huston-Tillotson at the forefront of AI education and ethical leadership is both bold and necessary. By anchoring an AI initiative in an institution rooted in history, excellence, and purpose, you show what inclusive innovation looks like. To the faculty, staff, and students of HTU: thank you for welcoming this work with openness, rigor, and pride. Your involvement affirms that the future of AI must be shaped by those who understand both its promise and its consequences.

To the HBCU AI Conference & Training Summit planning committee and participating HBCUs: Thank you for helping turn a vision into reality. In particular, I am grateful to Chris Hyams and Meme Styles for serving as co-chairs of the summit and leading with such care and wisdom. You set a standard for how academia, industry, and community can come together around responsible innovation. To all the Historically Black Colleges and Universities represented, and those preparing to join this movement, your engagement demonstrates that AI leadership must include all voices. HBCUs have always been engines of leadership, resilience, and innovation; your participation ensures that the AI era is no different. And to our allies across industry, nonprofits, and public service who stood with HBCUs as genuine partners, thank you for choosing collaboration over control, and for sharing access and influence to expand representation in tech.

Finally, to everyone else who has been part of this journey, whether through a conversation, a collaboration, encouragement, or challenge, you are present in these pages too. If you're reading this book and engaging with its frameworks, know that you aren't just applying a methodology. You are joining a movement.

Human First. AI Forward.
That's how we roll.
That's how we build

Introduction

We are living through a moment that will be studied for generations.

Artificial intelligence is no longer a future concept or a speculative technology. It is already reshaping how decisions are made, how work gets done, how value is created, and how power is distributed across organizations and society. Algorithms now influence hiring, lending, healthcare, education, marketing, logistics, and governance. Systems once designed to assist humans are beginning to guide them. And yet, amid all this technological acceleration, one uncomfortable truth keeps surfacing.

Most organizations are not failing at AI because of technology.

They are failing because of people.

Across industries, leaders are investing heavily in advanced tools, data platforms, and automation, only to watch initiatives stall, underperform, or quietly collapse months later. The technology works. The strategy looks sound on paper. The capital is committed. And still, results fall short. Teams resist adoption. Decision-making slows instead of speeds up. Cultures fracture under pressure. Execution lags behind ambition.

This is not a technology problem. It is a leadership problem.

The pace of innovation has outgrown the human systems responsible for leading it.

For decades, organizations optimized for efficiency, predictability, and control. Those models worked in a slower world. They rewarded specialization, hierarchy, and risk minimization. But the AI era is fundamentally different. It is defined by velocity, ambiguity, and constant change. In this environment, outdated leadership models do not just underperform. They actively break.

The question facing leaders today is not whether AI will transform their organization. That is already happening. The real question is whether their people, culture, and leadership systems are prepared to absorb and direct that transformation in a way that creates value rather than chaos.

This book was written to address that gap.

It introduces a new way of thinking about leadership, innovation, and execution in an AI-driven world. One that starts not with machines, but with humans. One that recognizes that technology is an amplifier, not a substitute, for human capability. And one that treats people not as a variable to manage after the fact, but as strategic infrastructure that must be intentionally designed from the start.

Throughout my work with executives, boards, investors, universities, and global organizations, I have seen the same pattern repeat. Leaders articulate bold visions for transformation, yet underestimate the readiness of their teams to execute those visions. They assume alignment where there is confusion, capability where there are skill gaps, and trust where there is unspoken resistance. When AI enters the equation, those hidden fractures widen rapidly.

I call this gap the Human Delta™, the space between strategic ambition and human readiness.

The larger the Human Delta™, the greater the risk that even the most advanced AI initiatives will fail to deliver their promise. And the faster technology moves, the more costly that gap becomes.

This book is about how to close it.

It is written for leaders who carry responsibility. Responsibility for people. Responsibility for decisions. Responsibility for systems that shape outcomes far beyond quarterly results. Whether you are a board member governing long-term risk, an executive leading transformation, an investor evaluating teams, an educator preparing future leaders, or a manager navigating change on the front lines, the challenges you face share a common thread. You are being asked to lead in conditions of uncertainty, speed, and complexity that previous generations did not encounter at this scale.

You are also being asked to do so with tools that are powerful, opaque, and evolving faster than most governance structures can keep up with.

This book does not argue against AI. On the contrary, it assumes AI will continue to advance and embed itself deeply into our institutions. The question is not whether to adopt AI, but how to lead it responsibly and effectively.

The central premise of this book is simple, but often overlooked.

The future will not be defined by who adopts AI first.

It will be defined by who leads it best.

Leading AI well requires a shift in mindset. It requires moving beyond optimization alone and toward humanization. It requires understanding how people actually innovate, collaborate, adapt, and respond to uncertainty. It requires designing teams intentionally, aligning leadership at every level, and building cultures that are resilient under pressure rather than brittle when stressed.

It also requires humility.

AI exposes misalignment. It magnifies cultural weaknesses. It accelerates both strengths and flaws. Leaders who view technology as a silver bullet often discover that it simply reveals what was already broken. Leaders who approach AI as a human systems challenge, however, gain a powerful advantage. They learn to anticipate friction, engineer balance, and turn diversity of thought into velocity rather than conflict.

This book provides the frameworks, language, and lenses to do exactly that.

You will explore how innovation truly happens inside organizations, not as a stroke of genius, but as a team sport shaped by complementary human strengths. You will learn how to make the invisible dynamics of innovation visible and measurable. You will see why balanced teams consistently outperform homogeneous ones, and how small changes in human configuration can unlock outsized returns.

You will also encounter real-world examples that demonstrate what human-first, AI-forward leadership looks like in practice. Not as theory, but as applied strategy. These examples are not meant to be copied wholesale, but to illuminate patterns you can adapt within your own context.

This is not a book that must be read linearly from beginning to end, although it can be. Different chapters will resonate differently depending on your role and responsibilities. Some readers may focus on governance and strategy. Others on culture, team design, or education. The frameworks are interconnected, but intentionally modular.

What matters most is not memorizing concepts, but changing how you see.

By the time you reach the final chapter, the goal is not simply that you understand AI better. It is that you understand leadership differently. That you see people as the primary lever of innovation. That you recognize culture as an engineered system, not an emergent accident. And that you feel more equipped to lead with clarity, confidence, and humanity in the face of accelerating change.

The AI era does not need less human leadership. It needs more of it, done better.

This book is an invitation to lead that way.

What This Book Will Change for You

By the time you finish this book, you will no longer think about artificial intelligence, leadership, or innovation the same way.

You will stop asking whether your organization is "AI-ready" and start seeing the deeper truth: success in the AI era is not determined by technology, but by human readiness. You will learn how to identify and close the gap between bold strategy and real execution, the gap I call the Human Delta™, before it quietly undermines your most ambitious initiatives.

You will shift from treating people as resources to be optimized to seeing them as strategic infrastructure to be intentionally designed. Instead of hoping the right culture emerges, you will understand how to engineer alignment, trust, adaptability, and innovation across teams at every level of the organization.

You will gain a new lens for decision-making. Rather than defaulting to speed, automation, or efficiency alone, you will learn how to lead with clarity, empathy, and purpose, using AI as an accelerator of human potential instead of a substitute for it. You will become more confident in navigating uncertainty, not by having all the answers, but by knowing how to ask better questions and design better systems.

You will also redefine what leadership looks like in this moment. You will move beyond outdated management models built for a slower, more predictable world and step into a human-first, AI-forward approach that is adaptive by design. You will learn how to build teams that balance vision with execution, creativity with discipline, and speed with responsibility.

Most importantly, this book will change how you see your role in shaping the future. You will recognize that leading in the AI era is not about keeping up with technology, it is about setting the human standard that guides it. Whether you are a board member, executive, investor, educator, or emerging leader, you will finish this book better equipped to lead change that actually works, because it works for people.

Artificial intelligence does not introduce new ethical questions; it amplifies the ones we have avoided for decades.

What happens next is not a technology problem. It is a leadership choice.

The organizations that will thrive in the AI era will not be the ones that move the fastest, but the ones that design human systems with intention. They will invest in judgment, trust, and capability long before a crisis forces their hand. They will understand that alignment is not accidental, culture is not self-organizing, and innovation is not sustainable without accountability.

This book does not offer shortcuts. It offers a standard. A way to think, decide, and lead when the stakes are high and the path forward is unclear. The frameworks that follow are meant to be used, tested, and refined.

The future of AI will not be decided by engineers alone. It will be shaped by boardrooms, classrooms, communities, and the everyday choices of leaders who understand that progress without humanity is not progress at all.

Human First. AI Forward. is the responsibility of leadership at this moment. And once you see it, you cannot unsee it.

PROLOGUE

❦

Human First, Leadership in Practice

A I won't break your company, but your leadership system will. This stark warning underpins the urgency of our moment. We stand at an inflection point in business and education. For the first time, the tools we've created are learning faster than we are. Yet the pressing question isn't whether artificial intelligence will transform the world, it's whether we will transform ourselves to lead it responsibly. In the race to adopt AI, most organizations focus on technology and data, but success or failure hinges on something far less technical and far more human: leadership.

Artificial intelligence is advancing at breakneck speed, reshaping industries and redefining how value is created. Every day brings new AI capabilities that can automate tasks, generate insights, and accelerate processes. Every executive feels pressure to "get ahead of the curve." But amid this frenzy, a critical truth is often overlooked: technology alone doesn't determine outcomes, people do. The greatest risk of the AI era isn't a rogue algorithm or a hardware failure; it's a leadership breakdown. Studies estimate that upwards of 80% of AI projects fail to meet their objectives, not due to faulty code or insufficient data, but because human systems, culture, teams, decision-making, were unprepared for the complexity and speed AI brings. In other words, when cutting-edge AI collides with outdated leadership habits, the initiative falters and human friction takes over.

This book is a rallying cry to flip that script. Human First. AI Forward. argues that in an age of intelligent machines, our competitive edge comes from elevating the people behind the machines. This is not a warning against innovation; it's a call for a new kind of leadership. The premise is simple: technology doesn't automatically create progress, leaders do. Your organization's AI strategy will only go as far as your people's capacity to understand it, embrace it, and unleash it. The systems we design reflect the values and biases of those who design them. If our leadership teams aren't equipped to lead with clarity, inclusivity, and purpose, then even the best technology will underperform or cause harm.

Consider Chris Hyams, former CEO of Indeed, one of the world's largest job platforms. He saw firsthand how algorithms were increasingly shaping economic opportunity for millions. After stepping down as CEO, Hyams chose to lead in a new arena, responsible AI education. He joined Huston-Tillotson University (a historically Black institution) as a board member and visiting professor to teach and mentor the next generation of AI leaders. Why? Because he recognized that if AI is going to transform work and society, then voices historically excluded from technology must be at the center of designing its future. Hyams understands that real innovation requires proximity to those impacted and diversity of perspective. By teaching AI in a community rich with perspectives often missing in Silicon Valley, he's helping produce leaders who treat ethics not as an afterthought but as part of the engineering discipline. His leadership lesson is clear: who builds and guides AI matters just as much as what we build.

Now consider Meme Styles, the founder and CEO of MEASURE, a nonprofit championing data equity. She starts from a simple premise:

communities most affected by technology should have a hand in shaping it. Styles' organization equips grassroots communities with tools to collect data, evaluate AI systems, and influence policy on issues like policing, healthcare, and education. The result? When the people on the ground help test and refine AI, biases are exposed earlier, trust increases, and the technology performs better in the real world. Styles demonstrates that ethical, human-centered AI isn't about slowing down progress, it's about expanding participation. By inviting more voices into the process, she creates better outcomes and greater buy-in. Her mantra proves true: progress that isn't inclusive isn't truly progress.

These leaders exemplify what I call the Human Standard. They prove that we can harness AI in ways that amplify human potential rather than replacing it. They reject the false trade-off between speed and responsibility, or between innovation and inclusion. Instead, they show that ethics influence performance, inclusion drives scalability, and community trust builds resilience. In short, when people are put first, technology actually works better.

The chapters that follow will equip you with a new leadership playbook for the AI era. You'll discover how to close the widening gap between technological change and organizational readiness, a gap I refer to as the Human Delta. You'll learn how to map your team's innovation DNA and re-engineer your culture to be deeply human, endlessly adaptive, and fearlessly innovative in the face of chaos. We'll delve into a groundbreaking framework (the Innovation Strengths Preference Indicator, or ISPI™) that makes the invisible dynamics of team innovation visible and measurable. And we'll explore real-world case studies, from a collegiate leadership society to a Fortune 100 company, all illustrating one core truth: organizations don't innovate, people do.

This is not a tech manual. It's a leadership manifesto for a world of intelligent machines. By the end, you'll have the key questions to ask, the metrics that matter, and the human-centric strategies to ensure your organization doesn't just survive the next wave of disruption, but defines it.

The future isn't about who adopts AI first, it's about who leads it best. Whether you're a CEO trying to transform a company, a university dean reimagining education, or an association director rallying an industry, the path forward is the same: Human First. AI Forward. That means leading with empathy, clarity, vision, and letting technology follow as an accelerator, not as an end in itself.

If you're ready to cut through the hype and lead with clarity amid the AI chaos, read on. The stakes could not be higher. This is the moment to re-architect how we lead, so that when the story of the AI era is written, it won't be a tale of humans sidelined by machines, but of humans rising to the occasion and steering technology toward a better future.

Human First. AI Forward. That is how we will build the future together.

"Technology does not break
organizations.
Leadership that fails to evolve does."

— **Greg Brisco**

CHAPTER 1

❧

The Velocity Problem, When Tech Outpaces Leadership

We are living through the fastest transformation curve in human history. Artificial intelligence isn't simply enhancing industries; it's rewriting them. Yet amid this rush to deploy AI everywhere, one dangerous truth is often overlooked: technology alone is not what breaks organizations. It's the human system leading that technology. In an AI-driven world, our organizational reflexes haven't caught up to our technological reality.

By now, surveys estimate that 70% - 85% of AI initiatives fail to deliver their intended value. Tellingly, these failures rarely happen because of flawed algorithms or insufficient data science. They happen because of people-related factors: misaligned teams, outdated management models, bottlenecked decision-making, and a fundamental lack of readiness for the complexity that AI introduces. When cutting-edge AI tools meet twentieth-century leadership habits, trouble follows.

I call this the velocity problem: the breakneck pace of technological change colliding with leadership capacity built for a slower era. In one recent study, 81% of business leaders said the biggest barriers to getting ROI from AI were people-related issues like culture and leadership, not the technology itself. Similarly, Gartner analysts project that roughly 85% of AI projects fail to meet expectations, primarily due to human factors rather than technical ones. In other words, AI itself won't wreck your company; the way your people lead and work together will.

The symptoms of the velocity problem are everywhere. Companies pour millions into AI projects, only to watch execution stall because teams aren't aligned or trained. We roll out powerful algorithms, only to encounter internal resistance and confusion. We invest heavily in data and models, but neglect the people who must implement and use them, and then wonder why adoption falters. The uncomfortable reality is that success or failure rests with people, not tech.

Traditional management frameworks were designed for stability and predictability. Hierarchies, silos, long planning cycles, these were playbooks for a world moving at 30 mph. Now we're driving 150 mph, and those old structures are starting to shake. AI brings volatility, uncertainty, complexity, and ambiguity on an unprecedented scale. Yet many leaders try to navigate 21st-century turbulence with the same command-and-control habits of the past. The result is a growing execution gap: organizations announce bold AI strategies but struggle to realize value because their operating norms haven't evolved.

It's tempting to blame technology when an AI project underperforms. But the truth is stark: success or failure lives and dies with people. For example, one global survey by NTT Data found that more than 80% of executives view workforce and leadership issues as the biggest barrier to achieving AI benefits. Likewise, when projects implode, it's often not because the tech didn't work, it's because teams fractured under pressure or leaders failed to adapt. In plain terms: AI won't break your company; your leadership system will. Unless we radically adapt how we align people, make decisions, and shape culture, the speed of change will continue to outrun leadership's ability to handle it.

Why do these fractures often go unnoticed until it's too late? Because in times of stability, misalignment and inertia can hide under the surface. A company running on legacy leadership practices might seem fine in a calm market. Things "work" well enough. But introduce a shock, say, an ambitious AI initiative, and it stress-tests every weakness. Suddenly, hidden cracks widen into chasms. Decision-making that was always a bit slow now completely bogs down when it needs to speed up. Siloed teams that barely communicated before now actively collide and block progress. A risk-averse culture that managed in calmer times now freezes in the face of AI-driven ambiguity, because employees were never encouraged to experiment or adapt. By the time missed deadlines and burnout reveal the problem, the damage is done.

Think of every organization as running on an implicit Leadership Operating System, the cultural code and processes that dictate how work actually gets done. In the past, an outdated "people OS" could limp along in a forgiving environment. In the AI era, its weaknesses are brutally exposed. Consider a company where decisions have always been top-down and slow. That might have felt orderly before, but now an AI project demands rapid, iterative decisions at the front lines. The old decision bottleneck becomes a fatal choke point. Or picture a leadership team where half the executives are bold innovators and the other half are cautious operators. In stable times they might politely ignore their differences, but under the stress of an AI rollout, that latent misalignment erupts into open conflict over direction and risk tolerance.

The velocity problem is, at its core, a leadership problem. It demands urgent attention from the top. As one advisor aptly put it, "Most organizations will not survive this curve, they are trying to solve 21st-century execution with 20th-century thinking." Unless leaders confront this mismatch head-on, even the best technical strategy will falter. The good news is that once we recognize the gap, we can begin to close it.

The rest of this book is about how to re-engineer our organizations for high-velocity change, how to build a leadership system that's as dynamic, adaptive, and intelligent as the technology now at our disposal. But first, we need a deeper understanding of the gap between intent and execution. We need to diagnose what I call the Human Delta, and why it's widening.

Key Takeaways (Chapter 1):

- Technology is outpacing leadership: The speed of AI-driven change is exposing a dangerous mismatch between modern technology and outdated leadership models. Many organizations are trying to meet today's volatility with yesterday's management playbook. This capacity gap is at the root of many AI failures.
- Most AI initiatives fail because of people, not tech: Roughly 80, 85% of AI projects don't deliver expected value, largely due to human factors rather than technical shortcomings. Misalignment, cultural resistance, skill gaps, and slow decision-making are the common culprits undermining AI efforts.
- Old habits break under AI pressure: Rigid hierarchies, siloed teams, and long decision cycles were built for a slower world. In fast-moving, ambiguous environments introduced by AI, those old habits lead to execution breakdowns. Hidden cracks in culture and process widen quickly under stress.
- The velocity problem is a leadership problem: When AI initiatives stumble, it's usually because the human systems (team design, leadership alignment, culture) weren't prepared for the pace and complexity. AI itself isn't usually the direct cause of failure, it's how people lead, collaborate, and adapt around the AI that makes the difference.

- Adapt or falter: Unless leaders proactively re-architect their organizations, aligning people, decisions, and culture to be far more agile and adaptive, the gap between technological potential and actual results will keep growing. The companies that thrive will be those that upgrade their leadership operating system to match the speed of innovation.

"The greatest risk in the AI era is not technological failure, but the gap between bold vision and human readiness."

— **Greg Brisco**

CHAPTER 2

❧

The Human Delta™, Closing the Execution Gap

In my years leading sales teams and global divisions through technological change, I noticed a pattern. The success of every new tool or system wasn't determined by the technology itself; it was determined by our people's ability to absorb and apply it. Again and again, I saw a troubling chasm between what leaders intended to achieve and what the organization was actually capable of executing. I call this chasm the Human Delta™, the gap between strategic ambition and organizational readiness. It's the silent execution gap between declaring "We're going to transform" and actually being able to do it.

In the age of AI, the Human Delta can quietly grow into a giant void. Leaders may declare bold, AI-first strategies, genuinely believing in a tech-driven future. But if their workforce isn't aligned, skilled, and psychologically prepared to work in new ways, those strategies collapse in practice. The larger the Human Delta, the greater the risk that big investments will fizzle out. Many organizations leap into AI projects with a significant Human Delta hidden beneath the surface. The gulf undermines innovation and often becomes glaringly apparent only 18, 24 months later, after budgets are blown and momentum is lost. By the time executives realize their grand AI plan is off-track, this unspoken gap between ambition and reality has already done its damage.

The Human Delta is essentially the blind spot where leadership's optimism disconnects from workforce reality. Consider a scenario: A CEO and board assume their company can seamlessly implement an AI-powered supply chain system in one year. They eagerly fund top-tier software and data integration. But on the ground, the teams lack the necessary data literacy; mid-level managers aren't comfortable with agile experimentation; front-line employees worry the AI will complicate their jobs or even replace them. The initiative launches into this unready environment and quickly sputters. The difference between leadership's aspiration and the team's true readiness, that's the Human Delta. AI didn't create the gap; it simply amplified the cracks that were already there.

And AI will widen that gap if not addressed. New technology has a way of exposing all the small misalignments in an organization and making them big. If communication was poor, an AI project creates more confusion. If decision-making was slow, AI's rapid pace creates chaos. If trust was lacking, AI changes spark fear and rumor. In short, any organizational weakness gets put on blast once AI enters the mix. This is why some companies find, a year or two into their AI journey, that they are worse off in execution than before, not because AI is inherently disruptive, but because it illuminated (and exacerbated) issues leadership wasn't addressing.

Recognizing the Human Delta reframes how we pursue AI or any major change. Instead of asking "Do we have the right technology?" savvy leaders also ask "Do we have the capacity and alignment to use this technology effectively?" If the answer is uncertain, that's a red flag. Addressing the Human Delta means systematically checking the human fundamentals: Do our people have the skills and training they need? Do they trust leadership and each other? Are incentives aligned with the change? Is our culture resilient and adaptable when setbacks occur? These aren't touchy-feely questions, they are predictors of whether expensive projects will actually deliver results.

One powerful example comes from a global company I advised during a digital transformation. The CEO rolled out an ambitious AI-driven customer service platform. Technically, it was brilliant, the system could predict customer needs and suggest solutions in real time. But six months in, customer satisfaction hadn't budged. When we dug in, the issue became obvious: the call center reps didn't trust the AI's suggestions. They ignored most of those recommendations, sticking to the old scripts they were comfortable with. Why? Because management hadn't invested in training them properly, nor addressed their fears about this new "AI assistant." The fancy technology was in place, but the Human Delta, a lack of trust, insufficient training, fear of the unknown, completely undercut its impact. We had to step back and close that gap with hands-on training sessions, feedback loops, and by involving the reps in refining the AI's recommendations. Only then did adoption improve and customer satisfaction start climbing.

The lesson is clear: No amount of AI can compensate for a low level of human readiness. A smaller, well-prepared team will beat a larger, disjointed one even with the same tech. The Human Delta reminds us that human capacity is the true limiting factor in innovation today. If you reduce that delta, by raising skills, aligning teams, and cultivating an adaptive culture, technology ROI skyrockets. Ignore the delta, and even the best tech will underperform.

So, how do we systematically close the Human Delta? First, by measuring it. Leaders need a clear view of where their organization stands on the factors that drive execution. This means assessing skills (e.g. digital literacy, innovation skills), gauging alignment (are middle managers and front-line teams on the same page as leadership?), and testing adaptability (how do teams handle change and uncertainty?). In later chapters, we'll discuss tools and metrics, an "innovation readiness scorecard" of sorts, to make these fuzzy areas more concrete.

Second, by treating change management not as a box to check at the end of a project plan, but as a core design principle from day one. That means involving people early and often, communicating the "why" behind the AI initiative, and empowering champions at all levels. It means anticipating where fear or confusion might arise and addressing it proactively. In short, it means leading with empathy and clarity at every step, so people come along on the journey rather than feel dragged into it. If employees understand the purpose of a new system, have a chance to voice concerns, and see their feedback shape the rollout, they're far more likely to adopt the change enthusiastically.

Finally, closing the Human Delta requires a mindset shift: from seeing people as executors of a plan to seeing them as co-creators of change. When employees participate in shaping how an AI tool will be used, their buy-in soars. When teams are invited to experiment and provide feedback, not only does the technology implementation improve, but the people grow more confident and invested. Leaders who shrink the Human Delta do so by engaging and developing their people at each stage, not just announcing a change and then commanding everyone to get on board.

As we move forward, keep the Human Delta in mind. In every case study and framework we explore, ask yourself: how does this help bridge the gap between what we want to achieve and what we're ready to achieve? By persistently asking that question, you'll train yourself to spot hidden misalignments before they sabotage your strategy. You'll start to see every ambitious goal in two parts, the technical challenge and the human challenge, and address both in equal measure.

In the AI era, the organizations that win won't necessarily be those with the fanciest algorithms or the largest data lakes. They will be the ones with the smallest Human Delta, where people and tech move in sync, and strategy and execution are tightly aligned. Our journey now turns to how to achieve that state, starting with rethinking what we optimize for and recognizing why human factors are emerging as the ultimate differentiator.

Key Takeaways (Chapter 2):

- The Human Delta™ = The Strategy, Execution Gap: The Human Delta is the gap between leadership's bold intentions and the organization's ability to deliver. In the AI era, ambitious visions often outstrip the team's readiness to execute, causing initiatives to stall despite ample investment.

- AI exposes misalignment: New tech doesn't automatically fix execution issues, it exposes them. AI projects tend to magnify any lack of skills, trust, or adaptability in the culture. Small misalignments (in training, communication, etc.) become big blockers under the stress of AI-driven change.

- Every project has a hidden people factor: When an AI initiative falters, the key question isn't "What went wrong with the tech?" but "Where weren't our people prepared?" In many cases, employees haven't been properly trained, informed, or motivated to embrace the new tools. Human readiness is often the real make-or-break.

- Close the gap by investing in people: A team's capacity can be improved. Leaders can measure current readiness (skills, alignment, adaptability) and then boost it through training, hiring for missing strengths, re-organizing teams, and building a more open, adaptive culture. Narrowing the Human Delta, even modestly, dramatically increases the odds of tech success.

- Change is co-created, not imposed: Shrinking the Human Delta means involving employees as partners in the change. Organizations that engage people early, listen to feedback, and iterate the implementation with their input see far higher adoption and better outcomes. People support what they help create.

"You can optimize yourself into
irrelevance.
Humanization is where the real
advantage begins."

— **Greg Brisco**

CHAPTER 3

❦

From Optimization to Humanization, Redefining the Edge

For decades, the business playbook was all about optimization. We optimized supply chains for efficiency, optimized code for speed, optimized workflows to eliminate errors. That made sense in a world where incremental gains and cost savings won the day. But in this new era defined by AI and rapid change, mere optimization has become a commodity. In fact, you can optimize yourself into irrelevance, making a mediocre product marginally faster or an outdated process slightly cheaper. The real competitive edge now lies elsewhere: in what I call humanization, amplifying the uniquely human factors of innovation like creativity, empathy, trust, and adaptability.

Optimization is about doing things right; humanization is about doing the right things in ways only humans can. AI is incredibly powerful at the former, crunching numbers, streamlining tasks, optimizing decisions based on predefined parameters. But AI can't define purpose, inspire trust, or think outside the box of its training data. Those are human domains. Leading companies today realize that to truly differentiate in an AI-saturated market, they must double down on the human qualities that machines cannot replicate.

Consider how customer experience has evolved. The old approach was to optimize every touchpoint for maximum efficiency, minimize wait times, reduce costs, automate responses. Important goals, yes. But the new approach is to humanize the experience, to empower employees to deliver empathy and personal connection, and to use AI behind the scenes to augment their ability to serve customers. The companies winning loyalty now aren't necessarily the ones with the fastest chatbot or the most polished app; they're the ones that make customers feel understood and valued, with AI working in the background to free up humans for real human-to-human moments. In other words, the standout companies might use the same AI as everyone else for basics, but they differentiate by layering a human touch on top.

We can feel this shift from optimization to humanization inside organizations as well. Traditional management asked, "How do we get more output from each employee?" a very mechanistic, efficiency-driven question. The emerging leadership mindset asks, "How do we unlock the full innovative potential of each person?" The first question treats people like cogs in a machine to be fine-tuned; the second treats people as creative engines to be unleashed. It's not that efficiency and productivity no longer matter, they absolutely do. But in a humanized model, high efficiency is a byproduct of engaged, creative, mission-driven people, rather than an end in itself achieved at the expense of morale or imagination.

Why is humanization becoming the decisive factor? Because in a world where every competitor has access to similar AI tools and data, how you apply them, the human context you build around them, determines success. Two firms might deploy the same AI for, say, financial portfolio management. One optimizes purely for short-term profit and speed. The other humanizes its strategy by also considering ethical implications and long-term relationships with clients. Over time, the second firm builds far greater trust with stakeholders and adapts better when the market shifts unpredictably (something the purely optimized algorithms fail to anticipate because they lack human judgment and values). The human-centric approach proves more resilient and sustainable.

Another illustration: during the COVID-19 pandemic, many organizations implemented AI-driven tools to monitor remote worker productivity. The "optimize at all costs" mindset led some companies to impose rigid digital surveillance, counting keystrokes, tracking active hours, and sending automated warnings for idle time. Technically efficient, perhaps, but disastrous for morale and trust. Forward-thinking leaders asked a different question: How can we use AI to support our people's well-being and effectiveness during a challenging time? They deployed tools that helped employees manage workload, facilitated better virtual collaboration, or identified who might be struggling and need a check-in. In other words, they humanized the application of AI. Those organizations emerged with stronger cultures and higher loyalty, whereas the companies that treated employees like machines faced backlash and turnover. The contrast was stark: one side optimized for output, the other optimized for the whole human, and the latter won in performance and retention.

To drive humanization, leaders must broaden the metrics of success. Traditional KPIs, output per hour, error rate, profit margin, tell a very narrow story. We need to also measure things like trust, learning, and engagement. Some progressive companies track a "Trust Index" among employees or run frequent pulse surveys to gauge whether people feel valued and heard. Others measure the rate of experimentation (e.g. number of new ideas tested per quarter) as a key metric, recognizing that a high experiment rate indicates a culture where people feel safe to innovate, an inherently human factor. These might sound "soft," but they directly predict hard outcomes like the innovation pipeline and agility in the face of change. If trust is low or people have stopped proposing ideas, that's an early warning of stagnation that no amount of AI investment alone can fix.

Crucially, humanization doesn't mean being "soft" or sacrificing results for feel-good culture. Quite the opposite: it's about recognizing that peak performance in an AI world comes from harnessing human creativity and collaboration to the fullest, which only happens in an environment of trust, purpose, and growth. Humanization is a performance strategy. Empathy can speed up innovation because teams that trust each other debate more openly and iterate faster. Inclusion yields better products because you avoid costly blind spots and design for a wider audience. Purpose galvanizes effort in ways that bonuses never will. These human elements are force multipliers for innovation and execution.

One powerful lever of humanization is storytelling and narrative. When launching an AI initiative, an optimization-focused approach might communicate only the technical specs or business case: "This new system will process claims 30% faster." A humanized approach builds a narrative around meaning: "By automating the tedious parts of claims processing, we're freeing you, our talented team, to focus on helping customers through the hardest moments in their lives, which only a caring human can do. This will make your work more meaningful, and our service more personal." Notice how the exact same project now feels completely different. The narrative taps into purpose and respects the human role. People are far more likely to embrace the change and make the most of the technology when it's framed this way, because they see how it elevates their significance rather than diminishes it.

As leaders, we have to consciously design human-first workflows. That means asking, at every implementation step: How will this make our people feel? Does this tool empower or surveil? Does this process encourage creativity or smother it? Are we using AI to enhance human judgment or to override it? These design choices determine whether AI becomes a friend or foe to your workforce. The best companies co-create answers to these questions with their employees. They involve teams in pilot programs and genuinely listen to feedback on how to make the tech more user-friendly and supportive. They iterate not just the code, but also the work environment around it.

Let's talk about mistakes and learning, because they're deeply human experiences. Optimization culture often treated mistakes as failures to be minimized or hidden. A humanized culture treats mistakes as fuel for learning. In high-performing innovative teams, a project setback or an AI model's error is openly discussed without blame, so the team can extract lessons and improve. When you humanize work, you cultivate psychological safety, people feel safe to say "I don't know" or "I messed up," which ironically leads to fewer critical mistakes in the long run because problems surface early and get fixed. Teams where people aren't afraid of repercussions will test, tinker, and push boundaries, exactly what you need for breakthrough innovation. By contrast, if everyone's terrified of making a mistake, all you get is cautious, incremental moves.

In summary, shifting from optimization to humanization means expanding our view of efficiency to include emotional and social dynamics, not just mechanical ones. It's recognizing that people aren't programmable components, and treating them as such yields only mediocre results. Treat people as creative, complex, purpose-driven beings, and support them with technology that amplifies those qualities, and you unlock something far more powerful than mere efficiency: you unlock engagement, passion, and ingenuity. The coming chapters will dive deeper into how to operationalize this philosophy. We'll look at how to redesign teams, leadership approaches, and cultures to be human-first and AI-empowered. But at every step, remember this core shift: optimization alone is no longer enough; humanization is the differentiator. The organizations that thrive won't be the ones that just run the machine a little faster, they'll be the ones that redesign the machine entirely around human brilliance.

Key Takeaways (Chapter 3):

- Efficiency isn't enough, human factors are the new edge: In an AI-saturated world, pure optimization (faster, cheaper, leaner) has diminishing returns. The true competitive advantage comes from humanization, leveraging creativity, empathy, trust, and other human strengths that technology can't replicate. Companies that foster trust, purpose, and diverse thinking will outperform those that only chase efficiency.

- Use tech to augment humans, not the other way around: A human-first design flips the typical question. Instead of "How can we use people to maximize our tech?", ask "How can we use tech to maximize our people?" AI should augment human abilities, not replace the human touch. Workflows and customer experiences designed with empathy and flexibility will beat purely algorithmic solutions in the long run because they drive deeper engagement and loyalty.

- Culture is a performance driver, not a soft afterthought: Psychological safety, inclusivity, and continuous learning aren't just "nice-to-have" values, they directly impact innovation and execution. When employees feel safe to take risks and share bold ideas, organizations avoid blind spots and adapt faster. In contrast, a fear-driven, metrics-only culture stifles the very human inputs (imagination, candid feedback) that fuel breakthroughs.

- Measure what truly matters (including "soft" metrics): Forward-looking leaders track indicators like trust levels, employee engagement, rate of experimentation, and customer delight, not just output and efficiency. These human-centric metrics provide early warning signs of whether your organization's innovation capacity is growing or eroding. If trust dips or people stop experimenting, no amount of AI investment will drive results until those issues are addressed.

- Lead with purpose and narrative: People respond to meaning. Framing an AI initiative with a clear "why", how it improves lives, helps employees, or furthers a mission, will rally far more support than a sterile ROI calculation. Every tech change is fundamentally a human change, and winning hearts and minds is essential. Use stories, values, and co-creation to turn skepticism into buy-in. In the long run, it's engaged humans, not just algorithms, that win the race.

"Every organization runs on an invisible operating system, and in the AI era, outdated leadership code crashes fast."
— **Greg Brisco**

CHAPTER 4

❦

Organizations Don't Innovate.
People Do.

How do you repair a leadership operating system built for a by-gone era? Not with small tweaks or hopeful pep talks. The truth is, nothing short of a fundamental overhaul will do. Instead of tinkering at the edges, we have to re-engineer the human side of innovation from the inside out. Traditional leadership development focuses on making individuals a bit better, a worthwhile goal, but no longer enough on its own. We need to dive deeper, down to how teams are composed, the cognitive diversity they contain, and the intrinsic problem-solving styles of our people. In short, we must reach the innovation DNA at the heart of the organization.

Innovation DNA Engineering™: A Blueprint for People-Driven Innovation

In practice, Innovation DNA Engineering™ means treating the design of teams and leadership structures as a true engineering discipline, grounded in data, science, and repeatable frameworks. As I often explain to clients, our goal isn't to run another leadership training workshop; it's to re-architect leadership capacity at the DNA level of the organization. We're not just offering strategic advice; we're mapping and re-coding the human building blocks of execution itself. It's a bold promise, but we can back it up because a proven system sits at the core of our methodology and consistently delivers results.

My team and I at Humanize Innovation learned early on a core truth: organizations don't innovate, people do. Every breakthrough, every game-changing idea, every successful project ultimately comes down to individual human beings or teams of humans working in the right way. Yet many companies still behave as if innovation can be decreed from on high or installed like a software upgrade. They spin up innovation labs, appoint a Chief Innovation Officer, run brainstorming workshops. Those efforts have their place, but they only scratch the surface. They rarely address the fundamental human architecture that actually drives innovation.

Think of innovation like building a high-performance race car. You wouldn't just give the driver a pep talk and hope for speed; you'd lift the hood and tune the engine. You'd balance the chassis and pick the right tires for the track conditions. Innovation DNA Engineering™ takes that same under-the-hood approach with human teams. We ask ourselves: Do we have the right components in our innovation engine? Is each team configured and tuned for the specific kind of innovation we need? If not, how do we redesign it?

One lesson stands out: innovation is highly contextual and deeply personal. Some people are big-idea generators, others are meticulous finishers. Some thrive amid ambiguity while others bring order to chaos. All are valuable, but the mix you need varies with the mission at hand. So rather than treating people as generic "resources," we profile each person's innovation strengths with scientific precision (more on that soon). Then we design team composition like a chemist formulating a compound, combining elements in just the right proportions to spark the desired reaction. If we're aiming for a revolutionary breakthrough, we'll assemble a very different team DNA than if we're pursuing an incremental process improvement. This deliberate matching of people to problems is the essence of engineering innovation capacity.

Another counterintuitive insight is that you usually don't need "better" people; you need to arrange your people better. Often the talent you require is already in your organization, it's just not configured optimally. We've seen companies packed with brilliant individuals still fail to innovate because those individuals were clustered in one silo or their cognitive styles clashed and canceled each other out. By contrast, we've watched modestly resourced teams punch far above their weight simply because they were deliberately composed with complementary strengths. In those teams, abilities fit together like puzzle pieces. It's a powerful reminder that innovation performance is an emergent property of a system, not just a sum of individual talents. To improve it, you have to redesign the system itself.

Shifting to an engineering mindset also means embracing measurement and data. Traditional HR might rely on opinion surveys or gut feel to gauge team health. In engineering mode, we turn to hard data, psychometric assessments, network analysis, execution metrics, to pinpoint exactly where the innovation process is breaking down. Maybe the data reveals a team has plenty of detail-oriented planners but almost no risk-takers, which would explain why bold ideas never get off the ground. Armed with that evidence, we can intervene surgically. Perhaps we introduce a "Pioneer" profile into the mix or coach the team to expand its risk appetite. This targeted approach is far more effective than one-size-fits-all training sessions or generic platitudes about "fostering innovation culture."

In short, organizations don't magically innovate on their own. People do. More specifically, it's the way people are engineered into teams and cultures that drives innovation forward. Once you accept that premise, the leadership mandate shifts fundamentally: it's no longer just about inspiring people to innovate, but about configuring people for innovation. Innovation becomes an engineering challenge. The key question becomes, How do we design the optimal human system to achieve a given innovation goal? That's the question Innovation DNA Engineering™ answers, and it's a game-changer. In the next chapter, we'll introduce a powerful tool that makes this approach practical, essentially an Ultrasound for mapping a team's innovation DNA with unprecedented clarity.

Key Takeaways (Chapter 4):

- Treat Team Design as Science, Not Art: Innovation DNA Engineering™ treats the design of teams as a science. Instead of relying on intuition or surface-level training, it focuses on deliberately structuring team composition and dynamics in a data-driven way to maximize innovation capacity.
- People Innovate, Not Organizations: No strategy or innovation lab will succeed if the right people aren't aligned and empowered. Real innovation improvement happens at the people level, deciding who's on the team, how they think and collaborate, and whether that fits the mission. Even the best ideas or technology will falter if the human pieces aren't properly aligned.
- Innovation as a System Property: Innovation success is an emergent property of a well-designed human system. The right mix of cognitive styles, risk tolerances, and talents can dramatically amplify creativity and execution. Conversely, a misconfigured team can squander even top-notch individual talent. In other words, great innovators need the right partners to truly shine.

- Data Over Guesswork: Moving from an "optimize people" mindset to an engineering mindset means using data and scientific tools to guide decisions. We don't guess what might be wrong, we measure it. We don't just encourage innovation, we architect the conditions for it. This evidence-based approach replaces guesswork with precision when building and fixing teams.

- Continuously Configure for the Challenge: Leaders should constantly ask, "Do we have the right human ingredients, arranged the right way, for this innovation challenge?" If not, it's a solvable problem. You can realign roles, add missing profiles, or restructure workflows, but only if you recognize team design as a key lever for success. Viewing talent as modular and configurable, rather than fixed, is a critical leadership skill in the innovation age.

"Innovation is not a mystery. It is a measurable
human pattern waiting to be
understood."
— **Greg Brisco**

CHAPTER 5

❦

Innovation Strengths Preference Indicator® (ISPI™), Mapping the Human Elements of Innovation

Innovation is often talked about as if it were an elusive art, sudden flashes of genius, creative sparks, lucky breakthroughs. But what if we could demystify innovation with the clarity of science? What if we could map the human factors of innovation the way a chemist maps elements on the periodic table? That's exactly the promise of the Innovation Strengths Preference Indicator® (ISPI™), a patent-ed assessment tool that gives us a high-resolution Ultrasound of how individuals and teams innovate. Unlike breezy personality quizzes or generic leadership tests that offer only broad strokes, ISPI™ delivers data-driven, actionable insight into how people create, collaborate, and execute. It reveals the hidden drivers behind innovative behavior.

ISPI™ was developed by Bob Rosenfeld, a chemist-turned-innovation expert who approached the challenge with scientific rigor. Rosenfeld's background in chemistry gave him an obsession with precision and repeatability. He refused to accept the notion that human innovation traits were "too messy" to quantify. Over 25 years of research, he me-thodically decoded the fundamental elements of innovative behavior. The result is a tool with remarkable depth and reliability.

To give a sense of its rigor: ISPI™ has a Cronbach's alpha (a statistical measure of consistency) of roughly 0.85, indicating very high test-retest reliability. Put simply, if someone took the assessment 100 times, they'd get essentially the same profile about 85 of those times. By comparison, many popular personality tests score in the 0.4 to 0.5 range, meaning results often swing with the person's mood or context. ISPI™'s scientific pedigree means it measures stable traits, the deep-seated preferences and tendencies that don't change on a whim or with office politics. That stability is crucial. It means leaders can work with solid, consistent data about their people's "innovation DNA," rather than vague, horoscope-like generalities.

So what exactly does ISPI™ measure? In essence, it maps a dozen hidden dimensions of human innovation capacity across three domains: cognitive, affective, and conative. That's a mouthful, so let's unpack it:

- Cognitive aspects cover how people think and process information.
- Affective aspects relate to how people feel, their emotions, values, and motivations.
- Conative aspects capture how people act, their instincts for taking initiative and executing tasks.

By examining all three domains together, ISPI™ draws a holistic portrait of an individual's innovation style. Most conventional tools only look at one facet, for example, a personality test might focus on cognitive style but overlook motivation or action orientation. ISPI™ covers all three, offering a far more complete and useful view of how someone approaches innovation.

Concretely, ISPI™ evaluates individuals on key spectra such as:

- Ideation vs. Execution: Does a person naturally gravitate toward generating novel ideas, or toward implementing and finishing projects? Innovation is like a relay race: you need visionaries to sprint with bold ideas (Pioneers) and finishers to carry those ideas over the finish line (Builders). Neither trait is "better", both are critical. A team full of dreamers will churn out concepts with no follow-through, while a team of only doers can execute efficiently but may lack bold new ideas. Knowing where each person falls on this spectrum helps avoid those pitfalls and balance the team accordingly.

- Risk Orientation: How comfortable is someone with ambiguity and risk? Some individuals are natural risk-takers who thrive on experimentation and bold bets. Others are more risk-cautious, preferring to methodically assess and mitigate before proceeding. Both orientations are essential in innovation. A team of all risk-takers might launch daring ventures but stumble into avoidable failures, whereas a team of all cautious planners might avoid disaster but never venture beyond incremental improvements. ISPI™ quantifies this trait so you can see a team's collective risk appetite and intentionally balance it. For example, if a team is overwhelmingly risk-averse yet tasked with disruptive innovation, you'll know to inject some bolder thinkers (and vice versa).

- Adaptive vs. Innovative Problem-Solving: This dimension reveals whether someone is more of an Adapter or an Innovator in their approach to challenges. Adapters prefer working within established structures, refining and improving what exists. They excel at optimization, making processes more efficient and reliable. Innovators, by contrast, love to challenge assumptions and pursue fundamentally new approaches. They're willing to break the mold or start fresh to find a breakthrough. Both styles have merit. Adapters ensure stability and continuous improvement, while Innovators drive transformation and leapfrog progress. A well-rounded team benefits from both.

Without Innovators, a group may get stuck in incremental mode; without Adapters, wild ideas might never be honed into practical solutions. By illuminating these tendencies, ISPI™ helps leaders manage potential friction (e.g., an Innovator vs. Adapter clash) so that it becomes productive rather than destructive.

- Preference for Structure vs. Ambiguity: Some people crave structure, clarity, and well-defined processes, they want a solid plan before acting. Others are comfortable diving into ambiguity and figuring it out as they go, iterating on the fly. ISPI™ measures this preference because it heavily influences how someone tackles innovation tasks. A high-structure person might excel at creating detailed project plans and setting clear criteria for success (great for later execution phases), whereas a low-structure person shines in open-ended discovery and rapid prototyping (great for early ideation). Neither extreme succeeds alone: too much structure too soon can smother creativity, while too little structure too late can derail implementation. Mapping where team members fall on this spectrum helps you sequence work and assign tasks to play to each person's strengths at the right phase. For example, during a free-form brainstorm, let the low-structure folks lead while the high-structure colleagues capture ideas and later organize them into an actionable plan.

- Collaboration Style (Soloist vs. Team Player): Does someone innovate best on their own or through collaboration? It's a myth that innovation is only about lone geniuses or only about group brainstorming, different people excel in different settings. Some individuals do their most creative thinking alone, preferring to mull ideas independently and develop a concept before sharing. Others thrive in a team setting, feeding off the energy of group discussions and building on others' ideas in real time. A healthy innovation process usually needs both dynamics: solitary deep work and rich team interaction. ISPI™ reveals who on a team leans toward solo ideation versus collaborative brainstorming. With that insight, leaders can make sure each person works in the environment where they excel. For instance, if one team member is a strong soloist surrounded by extroverted team players, you might give that person independent responsibilities where they can shine, and ensure group sessions don't inadvertently drown out their contributions. Likewise, you can coach the collaborative majority to create space for individual reflection so the quieter innovators have their ideas heard. The result is a team where both the "quiet genius" and the "people person" contribute at their best.

- Emotional Attitude Toward Innovation: Innovation is as much an emotional journey as a technical one. This category gauges traits like someone's tolerance for failure, resilience after setbacks, openness to critique, and general optimism vs. skepticism about new ideas. These affective traits heavily impact a team's innovation climate. For example, if a team is pushing boundaries on a risky project, setbacks are inevitable. Some people naturally treat failures as learning opportunities, they bounce back with optimism and keep experimenting. Others may take failure to heart, growing cautious or discouraged after a setback. ISPI™ makes these tendencies visible. Knowing who your resilient optimists are versus who your cautious evaluators are allows you to lead more effectively. You might pair a highly resilient team member with someone who gets discouraged easily, so the former can help lift the latter's spirits when experiments go awry. Or if an entire team scores low on openness to critique, that's a warning sign of a fragile culture, one where people might avoid hard conversations or ignore feedback, putting the project at risk. In that case, a leader could set new norms for open debriefs or bring in an outside facilitator to help the team learn how to give and receive constructive feedback. By surfacing these emotional and cultural factors, ISPI™ lets teams address them proactively, often turning what could have been a hidden liability into a managed strength.

All told, these dimensions (along with a few others that ISPI™ measures) combine to create each person's unique innovation profile, their personal "innovation DNA sequence." Importantly, the goal isn't to put anyone into a fixed box. On the contrary, it's to illuminate each individual's natural strengths and preferences when it comes to innovation, so they and their leaders can leverage those strengths intentionally.

For example, say someone's ISPI™ results describe them as "High Pioneer, High Risk-Taker, Low Structure, High Collaboration." Immediately, you have a nuanced picture: this person is likely a bold visionary who thrives in group brainstorming and improvisational work, but might struggle with rigid processes or detailed follow-through. It doesn't mean they can't execute or adhere to structure. Rather, it suggests those tasks will drain them unless they're paired with someone who has complementary strengths (perhaps a high-structure "Builder" type) or given tools and support to help them stay on track. The insight is a starting point for support and team design, not a limitation or excuse.

The real power of ISPI™ emerges when you aggregate these profiles for an entire team or even a whole organization. By combining individual innovation DNA profiles, we can essentially decode a team's collective DNA. And that is transformative. Just as a chemist can predict how certain elements will react together, we can foresee how a specific mix of people might collaborate, how decisions will likely be made, and where friction might arise.

For instance, if a team's composite profile reveals that everyone scores very high on Ideation but low on Execution, you can predict lots of exciting brainstorming and a real risk that projects will stall before the finish line. Similarly, a team made up entirely of Adapter types, people who favor incremental improvement, might get along smoothly day to day, but they could miss disruptive opportunities and be blindsided by more visionary competitors. Conversely, a team of all Innovators, who challenge every assumption, might dream up bold proposals but struggle to converge on a practical plan or neglect crucial optimizations along the way.

ISPI™ team data can also highlight critical gaps. Maybe a certain division's profile shows almost no strong risk-takers, everyone in the group is highly cautious. That suggests any high-risk, breakthrough project assigned to them will likely falter unless you inject some more adventurous thinkers from outside or otherwise compensate. On the flip side, another team might skew heavily toward risk-takers with very few risk-managers; that's a signal to introduce stronger governance or add a "steady hand" personality to avoid reckless moves.

Seeing these patterns is a paradigm shift for leaders. It's like turning on the lights in a dark room, suddenly you can see the furniture and navigate around obstacles, instead of stumbling over problems you didn't know were there. ISPI™ makes the invisible visible, giving teams a common language and framework to discuss their dynamics openly.

In fact, when teams first see their ISPI™ results, "aha!" moments abound. All those unspoken frustrations or misalignments they've been feeling now have names and numbers attached. A team leader might realize, "No wonder our meetings have been tense, I have two fiercely independent creative types reporting to a manager who craves structure and predictability. They've been talking past each other." With the data in hand, that leader can broker a better understanding. Maybe they agree on a more flexible project process that still gives the manager enough order without smothering the creatives' flow.

Another example from my experience: a cross-functional task force was dragging its feet on developing an AI prototype. Through ISPI™, we discovered that most team members had a very low tolerance for ambiguity. They were excellent planners who wanted clear requirements at each step. But this particular project was exploratory, there were no clear answers up front. As a result, the team kept getting stuck, waiting for a certainty that was never going to come.

The solution was twofold: we added a couple of people with high ambiguity tolerance to the team to champion an iterative, experiment-as-you-go approach, and we coached the existing members to loosen their need for complete clarity in the early stages. The project's pace immediately accelerated after this team re-engineering. Nothing about the technology had changed, only the human configuration did.

Now, you might be thinking, "This sounds useful, but an assessment alone doesn't change anything." Exactly right. Data by itself is passive, it's what we do with the data that counts. ISPI™ is the foundational diagnostic tool in our Innovation DNA Engineering™ methodology, but it's only the beginning. The real power comes from how we apply those insights to actually engineer high-velocity teams and adaptive cultures. We'll explore that engineering process in the next chapter.

In short, ISPI™ lets us replace guesswork with intentional design. Instead of tossing a random mix of people onto a critical project and hoping for the best, we can deliberately craft teams whose innovation strengths fit the mission. We can anticipate friction points and address them before they blow up an initiative. We can spot leadership gaps, say, an overly homogeneous executive team that leaves the company exposed in certain areas, and take proactive steps to fill them.

As we often say, "Make the invisible visible, before it breaks your execution." Having ISPI™ data upfront is like having an early warning system for human dynamics issues that would otherwise only be discovered in a painful post-mortem. Armed with these insights, leaders can intervene with precision. They can realign roles, adjust project scopes, set up targeted mentorship pairings, or even rethink the feasibility of an initiative given the current team makeup.

To appreciate how broadly ISPI™ can be applied, consider its use across a variety of organizational contexts. The beauty of focusing on human innovation DNA is that it's universally relevant. Whether you're running a startup accelerator or a Fortune 100 division, a university program or a private equity portfolio, innovation ultimately comes down to people. Let's look at how several different groups, which you may belong to, can use ISPI™ to unlock significant value in very practical (and decidedly non-"salesy") ways.

ISPI™ for Membership Organizations and Professional Associations

If you lead or participate in a membership organization, say an industry association, a chamber of commerce, a professional society, or an executive network, innovation is likely high on your agenda. Members look to these organizations for thought leadership and practical tools to stay competitive. Introducing ISPI™ in this context can be a game-changer and a unique member benefit.

Imagine hosting an "Innovation Capacity Workshop" for your association's member companies. Each participating team could take the ISPI™ assessment and receive a profile of their innovation strengths and blind spots. For example, a marketing professionals association might have individual members assess themselves to see how they contribute to innovation in their own companies. The association could then facilitate peer coaching sessions where members with different profiles share strategies, for instance, a naturally cautious member might discuss how they learned to support bolder initiatives, while a visionary member might share how they partner with detail-oriented colleagues to execute ideas.

ISPI™ can also elevate community innovation initiatives that associations often run, such as industry hackathons or collaborative challenges. These events sometimes suffer when all the volunteers have similar backgrounds or thinking styles, imagine an accounting association's hackathon where each team inadvertently ends up composed entirely of accountants, all approaching a problem in the same careful, methodical way. By using ISPI™ data when forming teams, you can ensure each hackathon team has a healthy mix of perspectives. For instance, pair a methodical accountant (great at careful execution) with a visionary tech entrepreneur from another member company, and maybe add an academic with a high tolerance for risk. Not only will the teams likely produce more creative solutions, but participants will also experience different innovation styles firsthand and learn from each other.

Membership organizations themselves need to innovate to stay relevant, and ISPI™ can help internally as well. Consider your association's leadership or board: if all members have similar backgrounds and temperaments, the board might be prone to groupthink and slow to pursue new ideas. Perhaps the board is full of venerable industry experts who are excellent Adapters, preserving traditions and ensuring stability, but lacks Innovators who challenge the status quo. Using ISPI™ to assess the board's collective profile could reveal this kind of imbalance. In response, the organization might intentionally recruit a couple of pioneering, change-agile directors or committee members to inject fresh thinking. In this way, the association practices what it preaches: ensuring it has the internal innovation DNA to serve its members in a fast-changing world.

In short, for associations and professional networks, ISPI™ elevates the innovation conversation beyond generic calls to "be more innovative." It becomes a personalized development tool that deepens member engagement. Members feel truly seen and supported in developing their innovation capabilities, a significant value-add beyond the usual newsletters and webinars.

ISPI™ for Corporate Innovation Hubs and R&D Teams

Many corporations establish internal innovation hubs, labs, or incubators, dedicated teams tasked with pushing new ideas, whether in product R&D, process improvements, or exploratory AI projects. If you're leading such a hub or sponsoring one as an executive, you know the pressure: these teams must deliver results that defy the status quo, often under scrutiny from a skeptical core business culture.

There's a paradox we see often. Companies assign their "best and brightest" to innovation teams, yet "best and brightest" is usually defined narrowly, perhaps all brilliant technologists, or all charismatic creative types. Without deliberate balance, even a well-funded innovation lab can become lopsided. I've seen labs composed entirely of visionary brainstormers (walls covered in Post-it notes) but lacking people who know how to prototype systematically and drive ideas to implementation. I've also seen the reverse: an R&D division full of meticulous engineers and project managers that executes flawlessly but rarely questions core assumptions or produces breakthrough ideas.

ISPI™ is a godsend for these corporate innovation teams. First, it brings rigor and structure to the art of team design. Instead of relying on gut feel, "Let's put Karen on the innovation task force; she's creative", leaders can use data to ensure true cognitive diversity. For example, if an innovation project is exploring a new AI application, you'll want a mix of profiles: a tech visionary to push the envelope, a practical customer-focused thinker to keep it grounded, and a detail-oriented planner to map the route to market. ISPI™ can iden-tify employees across the organization who embody these traits, often surfacing "hidden" innovators in departments you wouldn't expect. We frequently find untapped creative talent tucked in operational roles; once their profiles are revealed, they flourish when given an innovation assignment.

Second, innovation leaders can use ISPI™ to diagnose and fix team is-sues quickly. Imagine two innovation sprint teams: Team A is thriving, Team B is floundering. Traditional thinking might blame external fac-tors (Team B got a tougher project or less budget). But ISPI™ might reveal a deeper cause: perhaps Team B's profile shows almost everyone on it prefers harmony and consensus. They get along well, but they may be too polite and too similar in outlook, leading to groupthink and an aversion to bold ideas. They converge on "safe" concepts and stall out. Meanwhile, Team A might have a healthy mix of a couple of big thinkers and a couple of skeptics who push each other, resulting in truly novel ideas that still get vetted.

With those insights, the innovation hub manager can adjust Team B's dynamics. Maybe you add someone with a contrarian streak to spark debate, or you explicitly assign a "devil's advocate" role in meetings to ensure challenging questions get asked. Without ISPI™, you might misdiagnose the problem as a technical hurdle or just assume Team B needs to work harder, missing the real issue.

For larger R&D organizations, ISPI™ data can inform talent devel-opment and role placement. When you know an individual's innova-tion strengths, you can put them where those strengths matter most. A scientist who scores high on solo ideation might produce brilliant insights working independently, but struggle in a committee-heavy project. So you give them a degree of autonomy to do deep work and plug them into the team at key review points. Conversely, an engineer who scores high on team collaboration and adaptability might be the perfect person to serve as a liaison with end users during a prototype trial, since they enjoy the back-and-forth feedback and can translate between technical and non-technical groups.

Honoring these preferences leads to better outcomes and happier innovators. Nothing disengages a creative mind faster than being stuck in a situation that constantly works against their grain, imagine a free-spirited idea generator forced to write compliance reports all day, or a structured planner told to "just brainstorm, we'll figure it out later." ISPI™ helps avoid those misplacements.

When innovation hubs embrace this approach, they start delivering on their promise. Breakthroughs actually move forward and make it to market, and they do so more smoothly because the team was engineered to handle the full journey from concept to execution. These teams become role models within the company, proof that innovation isn't just about luck or a few star individuals, but about assembling the right human ingredients. Instead of being quirky outliers that struggle to justify their budget, they become the engines of the company's future growth.

ISPI™ for Private Equity and Venture Capital Firms

In the high-stakes world of venture capital (VC) and private equity (PE), success often hinges on picking the right teams as much as the right products or markets. Seasoned investors frequently say they bet on the people, not just the idea. In fact, studies show that VCs overwhelmingly consider the founding team one of the most critical factors in investment decisions.

Yet how do investors evaluate a team beyond gut feeling and résumé bullet points? This is where ISPI™ can provide a significant edge by bringing structure and data to the assessment and development of human capital in a startup or a portfolio company.

For venture capital, imagine integrating ISPI™ into your due diligence process. When considering an investment in a startup, you could (with appropriate permissions and sensitivity) have the founding team complete ISPI™ profiles. The results might reveal, for example, that the founding trio are all extreme Pioneers, brilliant at vision and rallying excitement, but lacking a "Builder" who's focused on operationalizing and scaling the business. That insight doesn't mean you walk away, but it does influence how you structure the deal or support the company. Maybe you earmark funds to hire a COO with strong execution skills, or you plan to spend extra mentoring time with them on building operational processes.

Conversely, you might find a startup team that's technically solid and execution-focused (lots of Builder profiles) but lacks a big-picture Innovator who can guide them toward a truly game-changing strategy. As an investor, you could pair that team with a visionary advisor or ensure they add a board member who pushes them to think beyond incremental improvements. Essentially, ISPI™ acts as a risk radar for team dynamics, highlighting blind spots such as "all vision, no execution" or "all execution, no vision" before they become catastrophic.

For private equity firms, the use case often comes after an acquisition, during the rapid transformation or scale-up of a company. PE firms typically move fast to install new leadership, set aggressive targets, and shake up operations. Applying ISPI™ to the acquired company's management team can provide a roadmap for why the company was under-performing and how to unlock its potential.

Suppose a PE firm buys a mid-sized manufacturing company that has stalled out growth-wise. An ISPI™ assessment of the top team might show, for example, that nearly all the executives are highly risk-averse

Adapters, fantastic at efficiency and maintaining steady operations, but collectively hesitant to pursue new products or markets. No wonder the company has been stuck. The PE playbook might call for injecting new talent, and ISPI™ pinpoints what kind of talent is needed: maybe a CEO or Chief Digital Officer who scores off the charts as a Pioneer and risk-taker, to jolt the culture forward. The firm can even use ISPI™ in hiring those roles, ensuring the new leaders truly have complementary profiles rather than more of the same.

Throughout the ownership period, the PE firm could re-assess the company's innovation DNA to track progress, say a year in, did the leadership team's collective profile become more balanced? This adds a concrete, people-focused metric to the usual financial and operational KPIs of the transformation. Traditionally, PE firms rely on financial incentives and rigorous management to drive portfolio performance. ISPI™ gives them an additional lever: actively managing the human dynamics that underpin strategic execution, ensuring the company doesn't just execute better under new ownership, but adapts and innovates better too.

VC and PE firms can even turn ISPI™ inward on their own organizations. Investment firms themselves must innovate (finding new markets, adjusting theses) and make complex decisions as a partnership. A venture firm might profile its partners and discover, for example, that they're all strong Adapters, great at pattern recognition for proven business models, but possibly too cautious or consensus-driven to bet on truly disruptive startups. That firm might decide to bring on a new partner known for more innovative, contrarian thinking to avoid missing the next outlier opportunity. Or a PE partnership might find they're full of aggressive risk-takers (which is often the case in that field) and use that insight to institute stronger risk management protocols around decisions, or hire a more cautious operating partner to provide balance.

These may sound like subtle tweaks, but they can lead to significantly better decision-making and returns. In summary, ISPI™ allows investors to go beyond the surface level ("great team" vs "weak team") and quantify how a team is great or weak, and what exactly to do about it. In an industry where small execution missteps can make the difference between a 10x return and a total loss, that kind of foresight is worth its weight in gold.

ISPI™ for Incubators and Accelerators

Startup incubators and accelerators are unique innovation crucibles. They bring together early-stage entrepreneurs, often first-time founders, and aim to rapidly boost their growth through mentorship, structure, and resources. Anyone who has run an accelerator knows that startup teams vary widely: some soar, others stumble, and team dynamics often make the difference.

Accelerator programs usually emphasize business models, pitch coaching, and lean startup methodology. Those are vital, but there's often an overlooked question: what about the founding team's innate innovation strengths and gaps? Two startups might enter an accelerator with equally promising ideas, yet one thrives while the other falters, largely because one team gelled and the other was internally misaligned.

By integrating ISPI™ into an accelerator, you give startups a mirror to examine themselves early on. Think of a typical founding team in a tech accelerator: maybe a CEO with a grand vision, a CTO building the product, and perhaps a co-founder handling marketing or operations. They're heads-down on product development and investor pitches; rarely do they pause to discuss how their different working styles might complement or conflict.

If each founder takes the ISPI™ assessment, the accelerator can facilitate a candid, constructive discussion around the results. For example, suppose Founder A's profile shows a high Innovator score (loves to pivot and change things) and Founder B's profile shows a high Adapter score (prefers to refine an existing plan). That's a recipe for future tension if unaddressed. In a debrief session, the program director or a coach can say: "Let's talk about how you will handle decisions when a major pivot is on the table. You (Founder A) will naturally want to zigzag and try bold moves, whereas you (Founder B) will lean toward sticking to the plan. How will you manage that when the pressure is on?" These conversations can surface and resolve potential conflicts before they become irreparable rifts.

We've seen startups where, a year after the accelerator, the co-founders parted ways citing "we couldn't agree on the company's direction." In many cases, that kind of falling-out boils down to unrecognized style clashes, maybe one founder wanted to take big risks while another was more cautious. ISPI™ provides a neutral, nonjudgmental vocabulary for these differences. It's almost like team therapy for startups, but in a pragmatic business context. Each founder gets to articulate their perspective as a strength, and they learn how to value the other's approach rather than see it as an obstacle.

Accelerators can also use ISPI™ data to tailor mentorship. It's common to assign each startup one or more mentors, but not all teams need the same type of mentor. Say a particular startup team is brimming with technical genius and big ideas, a lot of Pioneer energy, but they lack operational discipline. The accelerator might pair them with a mentor who's a seasoned operations executive, someone with strong Builder traits who can teach them about process and execution rigor. Conversely, consider a team of brilliant engineers who are all Adapter types: they excel at incremental improvement on their current product, but may not be pushing imaginative leaps. That team could benefit from sessions with a mentor who's a bold Innovator, someone who challenges them to envision something radically bigger.

Without ISPI™, an accelerator director might not catch these nuances until it's too late, perhaps after a disappointing demo day or a missed development milestone. With ISPI™ profiles in hand from the start, you can customize the program for each startup. You'll have a better sense of which advice will resonate with which team, who might need extra support in certain areas, and even which startups might learn from each other.

That brings us to another powerful use of ISPI™ in accelerators: peer learning across the cohort. If you have, say, ten startups and they all take ISPI™, you can (with their consent) share some aggregate insights to spark collaboration. You might find, for instance, that Startup X and Startup Y are polar opposites in style. Perhaps Team X is fantastic at meeting every deadline (high structure, strong planning) while Team Y is brilliant at pivoting creatively when something isn't working (high adaptability). By highlighting these differences as complementary strengths, you can encourage them to exchange notes. "Team Y, talk to Team X about how you organize and hit your timelines. Team X, listen to Team Y about how you decide when to pivot."

This approach fosters a community where teams aren't secretly comparing themselves ("Why aren't we as creative as they are?"); instead, they're openly learning from each other's unique innovation DNA. It sets a tone that there's no single "right way" to innovate. Success comes from understanding and leveraging your own configuration, and knowing when to borrow a missing ingredient from someone else.

That lesson is invaluable as founders graduate from the accelerator and continue building their companies. They carry forward a mindset of self-awareness and appreciation for diverse working styles, which can make their future teams all the stronger.

ISPI™ for Higher Education Institutions

Colleges and universities might not be the first setting you think of for an innovation assessment tool, but consider the big push in academia around innovation and entrepreneurship in recent years. Campuses are creating innovation centers, interdisciplinary project courses, hackathons, and on-campus incubators. Whether you're an administrator, a professor, or a student leader, ISPI™ can be a powerful addition to your toolkit.

Take entrepreneurship and innovation courses, for example. Many universities have entrepreneurship minors or MBA programs where students form teams to launch startup ideas or tackle real-world problems. How are those teams usually formed? Often randomly, or by friends grouping together, or via a quick professor assignment. Rarely is much thought given to each team's internal dynamics. By introducing ISPI™ at the start of such a course, educators can help form more balanced teams or at least make teams aware of their collective strengths and weaknesses.

Imagine a team of four students discovers through ISPI™ that all of them are highly collaborative, big-picture thinkers, and none are detail-oriented or inclined to play the devil's advocate. An instructor armed with that insight might challenge them: "Who will play the skeptic or taskmaster in your group? Perhaps one of you needs to consciously step into that role, or you might seek input from someone outside your team who is more detail-focused." Suddenly, what could have been a stumbling block becomes a learning opportunity. The team can plan around their gap, and in the process, each student learns about leadership, self-awareness, and adaptation.

Furthermore, students gaining insight into their own innovation profiles can shape their career decisions. A student who learns, "I have a strong preference for execution and risk mitigation," might realize they find joy in roles that focus on implementing and scaling ideas. They might gravitate toward operations or project management roles within innovative companies, rather than feeling pressure to be the visionary founder (despite all the campus buzz around being an entrepreneur). Another student who shows up as a high Pioneer and high risk-taker might confirm that they're well suited to founding a startup or driving R&D in a cutting-edge company, and just as importantly, they'll understand that they'll need to partner with some strong executors to turn their bold ideas into reality.

University career counselors could also use ISPI™ insights to advise students. They can discuss how a student's innovation strengths align with different industries or job functions. For instance, a highly adaptive, big-idea person might thrive in a fast-moving tech startup environment, whereas a methodical, improvement-driven person might excel in a continuous improvement or quality management role at a large enterprise.

And let's not forget the faculty and staff. Universities often struggle with silos between departments, and interdisciplinary projects can flounder simply due to clashing work styles. Picture an initiative to launch a new AI research center that involves a computer science professor, a business school professor, and a dean. The computer science faculty member might be an Innovator and a bit of a lone wolf; the business professor could be more of an Adapter who values collaboration; the dean might be very high-structure and risk-cautious. Even if they all agree on the project's importance, their differing approaches could create friction. A facilitated ISPI™ session for the team could surface these differences in a neutral way. The group might realize, "Our computer scientist will chafe at long committee meetings. Let's allow him to work independently on a prototype while we (the committee) focus on broader parameters. Our business professor loves collaboration, so she can act as a bridge, synthesizing everyone's input.

And our dean is great with structure, so let's have her keep us grounded with timelines and reality checks." By acknowledging each person's style and role, the team can operate far more smoothly, instead of unknowingly pulling in opposite directions.

For universities committed to building a culture of innovation, using ISPI™ can improve outcomes on multiple levels. Students in project teams deliver stronger results and learn more from the experience. Research initiatives run by faculty hit their targets more often and with less infighting. Perhaps most importantly, everyone involved, students, professors, administrators, becomes more skilled at teamwork and self-aware leadership. That aligns perfectly with the broader mission of education: to unlock human potential. In this case, the potential we're unlocking is the ability to innovate effectively, supported by an understanding of the underlying "code" that drives each person.

ISPI™ for Executive Search and Leadership Placement

When companies seek new senior leaders, a CEO, a Chief Innovation Officer, a CTO, or any key executive, they often engage executive search firms to find the right fit. Traditionally, headhunters evaluate candidates based on experience, past performance, and interview impressions of "culture fit." In recent years, some search firms have started using behavioral assessments to avoid costly mis-hires. After all, placing an executive who looks perfect on paper but then clashes with the company's culture or fails to drive the needed changes can set an organization back significantly.

ISPI™ can take executive search to the next level. Consider incorporating ISPI™ into the vetting of top candidates. This isn't about disqualifying someone based on a profile score; it's about adding another critical data point to inform the decision. Suppose a company is hiring a Chief Innovation Officer tasked with reinventing its product line. The search firm identifies a candidate with stellar credentials, great track record, glowing references. Now imagine the ISPI™ assessment reveals this candidate's innovation DNA: perhaps they are heavily an Adapter, very structured, and somewhat risk-averse. In other words, they excel at optimizing and refining, but they may be less comfortable championing disruptive ideas. If the company truly needs breakthrough innovation, that profile could signal a mismatch.

Meanwhile, another finalist might show a profile of strong Pioneer tendencies, high risk orientation, and high resilience. That combination could be much better aligned with a mandate to shake things up. The goal here isn't to pick one candidate over another by numbers alone, but to ensure the person chosen actually fits the challenge and complements the existing team.

Sometimes, ISPI™ might highlight that if the company's current leadership team is ultra-cautious, bringing in a bold risk-taker is exactly what's needed, provided everyone understands and supports that new dynamic. Conversely, if the culture is already freewheeling and chaotic, perhaps the best hire is someone with a more structured, Adapter profile to bring balance. ISPI™ enables search firms to articulate these nuances clearly to their clients. For example: "Candidate X has the technical skills and experience you're looking for, and her innovation profile suggests she'll naturally complement your executive team by bringing a strategic risk-taking ability that the team currently lacks."

This kind of insight can also extend into onboarding and development after the placement is made. The best executive search firms don't just fill a seat; they help ensure the new leader succeeds. If the newly placed executive and her team all have ISPI™ data, an HR leader or executive coach can use it to accelerate assimilation. Imagine a new VP discovering that her direct reports collectively score very low on risk-taking. Armed with that knowledge, she can immediately take steps to build psychological safety and encourage small, low-risk experiments to gradually raise the team's comfort with taking chances. Without that insight, she might have pushed her team too hard too fast, or grown frustrated by their cautious pace.

The opposite scenario could also happen: maybe the team is full of aggressive innovators, and the new leader is comparatively more cautious and pragmatic. Instead of feeling pressure to match their style, she can lean into her role as the voice of reason, knowing it's a strength the team needs to avoid reckless decisions. The ISPI™ profiles essentially give a new executive a user's manual for her team, and vice versa.

From the search firm's perspective, offering this level of analysis is a differentiator. Imagine you're a CEO client and your search partner presents not just résumés and gut feelings but also a thoughtful assessment: "Candidate A tends to lead through collaborative vision and consensus-building, while Candidate B is more of a decisive, action-oriented leader. Here's how each approach aligns with what we understand about your company's needs and culture." In a data-driven era, that value-add is powerful. It signals that the search firm is not just finding talented people, but matching the deeper leadership DNA to the role and organization.

Additionally, ISPI™ can support succession planning, which many executive search and leadership advisory firms assist with. Say you're helping a company identify internal candidates for a future CEO role. Often, succession planning is a somewhat subjective process based on current performance and maybe a general leadership potential rating. By adding ISPI™ into the mix, you might spot a rising executive who hasn't had an obvious "innovator" role yet, but whose profile reveals a big-picture Pioneer with high adaptability, just the kind of latent capability needed to steer a transformation. On the other hand, you might find that a beloved senior VP, though great in their stable current role, has an extremely low tolerance for ambiguity. If the CEO position will demand navigating high uncertainty, that's a clear development area for that VP or a sign that the company might need to look externally when the time comes.

In all these ways, bringing ISPI™ into executive search and placement makes the process far less of a guessing game. It becomes more of an evidence-based alignment exercise: matching the right leader to the right context and then smoothing their integration. For organizations, it means a higher success rate with new leaders. For the search firm, it means a reputation for thoughtfulness and more placements that stand the test of time, which is the ultimate win-win.

A Unified Language and a Human-Centered Lens

Across all these sectors and use cases, from member organizations and accelerators to corporate teams and boardrooms, ISPI™ provides a common language for innovation. It translates the fuzzy idea that "people are our greatest asset in innovation" into tangible profiles and concrete discussion points. In doing so, it ensures that human factors are not an afterthought or a nebulous concern, but a central part of innovation strategy.

This is not just a semantic shift; it is foundational to Innovation Readiness™, making sure people and teams are truly prepared to drive innovation when it counts.

Having a shared language for innovation strengths is empowering and inclusive. Leaders and team members can move beyond vague labels like "she's creative" or "he's not innovative," which are often unhelpful judgments (and sometimes clouded by bias), to more precise descriptions. For example, "She has a strength in adaptive execution and risk management, which we need to balance our team," or "He thrives in generating bold ideas but will benefit from support to drive them to completion." That level of clarity removes the personal sting from differences. It becomes less about one personality versus another and more about roles, fit, and how distinct strengths complement each other toward a common goal.

It's also important to note that ISPI™ isn't about pigeonholing anyone. Human beings are wonderfully complex and adaptable. In fact, once people become aware of their own innovation preferences, they often find it easier to stretch beyond their comfort zones when needed. We see this repeatedly. Someone receives their ISPI™ report and it feels like a lightbulb turning on. "This description nails me. No wonder I struggle in long planning meetings, now it makes sense. I'm a high spontaneous ideator. Knowing that, I can prepare differently for those meetings or partner with a colleague who helps keep me on track." Self-awareness is the first step toward growth, and ISPI™ delivers that self-awareness in the specific context of innovation work.

For leaders, it also provides situational awareness of the collective mindset and emotional tone of their teams, enabling more thoughtful, intentional leadership in moments that matter most.

As we wrap up this deep dive into the ISPI™ tool and its methodology, take a moment to reflect on how these insights might apply in your world. Perhaps you recognized yourself or your team in some of the profiles and examples. Maybe you lead a visionary group that could use a dose of practical execution discipline, or a team of reliable executors who might benefit from an injection of bold imagination.

The key takeaway is this: innovation can be engineered. By understanding the human ingredients at a granular level, we can intentionally design better recipes for success.

In the next chapter, we'll shift our focus to exactly that, how to take ISPI™'s diagnostic power and apply it to engineering high-velocity teams and organizations. We'll explore why this approach must be championed at every level, from the boardroom to the front lines, and how it plays out for the Board, the CEO, the CHRO, technology leaders, managers, and beyond.

Key Takeaways (Chapter 5):

- Innovation DNA: The Innovation Strengths Preference Indicator® (ISPI™) is a scientific tool that maps the hidden dimensions of how individuals and teams innovate. It provides a high-resolution "Ultrasound" into innovation-related traits, showing how people generate ideas, handle risk, solve problems, collaborate, and deal with ambiguity, far beyond what a résumé or generic personality test can reveal.
- Scientific Rigor and Reliability: Developed with a chemist's precision, ISPI™ is highly reliable (around a 0.85 Cronbach's α) and validated through extensive research. It's patented and has been proven in demanding environments from Fortune 100 companies to military units and universities. Unlike trendy quizzes that might fluctuate with someone's mood, ISPI™ measures stable preferences across cognitive, affective, and conative domains. Leaders get consistent, data-driven insight into their people's innovation DNA rather than fleeting or surface-level data.
- Balanced Teams, Visible Gaps: ISPI™ identifies critical innovation roles and tendencies, like visionary "Pioneers" versus practical "Builders," bold risk-takers versus cautious risk-managers, paradigm-breakers versus optimizers, solo thinkers versus team brainstormers, and more. By aggregating individual profiles, leaders gain a clear view of a team's overall makeup. Imbalances or hidden gaps that were once invisible become visible. This enables proactive alignment: if

a team has too much of one type and not enough of another, you can address it before it undermines a project.

- From Data to Action: Simply having data doesn't create change, using it does. ISPI™ insights enable evidence-based action, from reshuffling team roles and pairing complementary styles to providing targeted coaching and aligning tasks with people's strengths. When individuals understand their own innovation profile, they gain self-awareness to adapt and collaborate more effectively (making the invisible visible in daily work). When leaders understand the team's profile, they can intentionally engineer the human side of innovation. It's a radical improvement over guesswork and one-size-fits-all team building.

- Broad Applicability and Value: Every part of the innovation ecosystem can leverage ISPI™. Membership associations can help member companies and professionals pinpoint and develop their innovation strengths, making industry collaborations more productive. Corporate innovation labs and R&D teams can be assembled and managed more systematically, turning innovation from an art into a repeatable process. Investors (VCs and PE firms) can de-risk their bets by assessing and balancing the human capital of startups and portfolio companies, often catching team issues early. Incubators and accelerators can mentor startups more effectively by tailoring support to each team's style and preempting co-founder clashes. Higher education programs can better prepare students and faculty to innovate in teams by building self-awareness and fostering complementary partnerships. Executive search firms can improve leadership placements and success rates by matching a leader's innovation profile to an organization's needs and culture. In short, from boardrooms to classrooms, everyone benefits when human innovation DNA becomes part of the conversation.

"The future belongs to teams designed
for difference, not consensus."
— **Greg Brisco**

CHAPTER 6

❧

Engineering High-Velocity Teams, An Everyone, Everywhere Imperative

In the last chapter, we saw how ISPI™ shines light on the invisible dynamics of innovation talent. But data and insight are just the beginning. The real impact comes from applying that understanding to build high-velocity teams and organizations, teams that can consistently out-innovate and out-execute the competition. This is not a niche HR exercise or a feel-good team-building activity; it's a strategic imperative that reaches the very top of the organization and extends to every level of the workforce. In fact, the conversation about aligning people for innovation belongs in the boardroom and needs to cascade all the way to the newest employee on the front line.

Why should this be a boardroom issue? Because in an era of digital disruption and constant change, a company's ability to innovate and adapt has become fundamental to its market value and survival. Boards of Directors today worry about technological disruption, upstart competitors, cybersecurity threats, regulatory shifts, you name it. And in each of these areas, the common denominator of a successful response is the human element. The smartest strategy or the largest technology investment will falter if the right people aren't in place, configured, and motivated to execute it and keep adapting it.

There's a sobering statistic often cited: roughly 70% of digital transformation initiatives fail to meet their objectives, with cultural or human factors being a primary cause. In other words, a board can approve a multimillion-dollar innovation project, but if they're not also examining the human dynamics that will drive that project, disappointment is the likely outcome. As one report put it, "technology isn't the problem; culture is."

So if you're a board member, you should be asking: Do we have a clear view of our organization's innovation capacity? Do we know whether our top team has the right mix of forward-thinkers and pragmatists, risk-takers and risk-mitigators, creative visionaries and operational executors? If not, why would we expect our ambitious innovation strategy to succeed? You might discover a significant gap between your strategic ambitions and your people's readiness to realize them.

In this chapter, we'll break down how Innovation DNA Engineering™ becomes a guiding practice for every key role in an organization. We'll explore what it means for the Board, the CEO, the CHRO and people leaders, technology chiefs, strategy executives, mid-level managers, and every employee. The tone here will remain candid and pragmatic, a conversation about why these ideas matter to you in your role, without any hype. The goal is to offer insight you can use, to help you see yourself and your challenges in this narrative, and perhaps to discover a new approach to solving them. Let's start at the top of the house and work our way down, understanding why aligning human innovation strengths with strategy truly is everyone's business.

Innovation as a Boardroom Priority

Boards are increasingly expected not just to provide oversight, but to ensure the organization is future-proof. Traditionally, a board's agenda centered on financial performance, compliance, and selecting or advising the CEO. The best boards now realize they must also monitor strategic talent and culture risks, chief among them, the risk of an innovation vacuum. A company can have plenty of cash and a spotless compliance record, yet still fail if it cannot adapt to a changing world. Think of iconic companies that declined over the years: it usually wasn't one bad quarter that did them in, but an inability to innovate, often rooted in a uniformly cautious culture at the top or a blind spot in leadership perspective.

A forward-thinking board will ask pointed questions about human capability: "Do we have the talent and culture to execute our strategy? If our strategy calls for bold innovation, do we actually have bold innovators in key roles? If our environment is turbulent, do we have resilient, change-agile leaders and teams to navigate it?" These questions are just as critical as reviewing financial projections.

Bringing ISPI™ data into the boardroom gives directors a new lens on the company's strengths and vulnerabilities. It's like a dashboard for the organization's innovation health. Suddenly, the board can discuss topics like: Where might the next leadership clash come from? Are we too one-dimensional in our thinking at the top? For instance, a board might review an annual talent report augmented with ISPI™ insights and learn that among the top 50 executives, the vast majority share a highly structured, risk-averse decision style. That's a glaring red flag if the industry is being upended by agile startups, it suggests a systemic bias against taking the very risks needed to reinvent the business. Armed with that knowledge, the board can press management for a plan to inject more innovative thinking into the leadership ranks. This could mean recruiting a few outsiders with different profiles, launching targeted development programs for rising leaders, or even bringing in advisory board members who think outside the current corporate box.

I encountered a real scenario like this: The board of a financial services company realized, through this kind of analysis, that their entire

C-suite "looked" the same in terms of innovation style. They were all excellent operators, consistent, reliable, detail-oriented executives. That homogeneity explained why the company ran like clockwork yet was always late to market with new offerings. Armed with that insight, the board made it a priority in the next CEO succession and other top hires to bring in people with different innovation DNA. In effect, they said, "We have enough operators; we need a visionary or two." Over the next couple of years, that shift at the top cascaded into bolder strategies and a more agile culture throughout the company.

Boards can also use this human-centric approach to mitigate leadership risk. Think of it as succession planning through an innovation lens. If the CEO got hit by the proverbial bus, who is the backup, and would that person's innovation DNA suit the challenges the company will face? Or suppose the company needs to undertake a major transformation, like moving to a platform business model or embedding AI into its core operations. Does the current leadership team have what it takes to champion and execute that change? A board shouldn't micromanage these questions day-to-day, but by having an informed view, it can hold the CEO and CHRO accountable for actively addressing them.

In essence, a board that focuses on innovation capacity is treating it as seriously as financial capacity. When directors speak to shareholders or stakeholders about the company's future, they can confidently talk not just about market share or patent counts, but about the people who will drive the next chapter and why those are the right people for the job. In tech-driven sectors especially, that kind of narrative, demonstrating that the organization has the human DNA for innovation, is increasingly important for investor trust.

There's also an inward angle: boards might apply these principles to their own composition. After all, a board is a team, too. Board effectiveness benefits from cognitive diversity just like any other team. If every director has a similar background and temperament, the board is prone to groupthink and could miss early warning signs or unconventional opportunities. An ISPI™-informed board assessment might reveal, for example, that while the board has plenty of financial and legal expertise (often necessary), it lacks voices with a disruptor or innovator profile, say, directors who are technologists or entrepreneurial thinkers who naturally think outside the box.

Some progressive boards have started to recruit one or two directors specifically known as "innovation champions", people who will ask the provocative questions and challenge comfortable assumptions that management might not challenge on their own. Knowing the board's collective innovation DNA can guide this recruitment. It helps ensure the board itself stays dynamic and forward-thinking, rather than becoming a complacent overseer.

In short, when innovation becomes a board-level conversation, the entire company gets the message that adaptability and creativity are strategic priorities, not just slogans. That tone from the top empowers CEOs, executives, and employees alike to lean into innovation, knowing it genuinely matters to those governing the organization.

The CEO's Mandate: Shaping Culture and Strategy with Innovation DNA

If the board sets the tone, the CEO amplifies it into action. For a CEO, aligning innovation capacity with strategy isn't a "nice-to-have", it's central to delivering results in the modern era. Ultimately, the CEO is responsible for performance, and in an AI-driven, fast-changing world, success hinges on doing new things (and doing them fast), not just squeezing efficiency out of the old things. So why should a CEO care about ISPI™ and the human innovation DNA of the organization?

Because it provides a blueprint for how to actually achieve the bold goals in the strategic plan.

First, culture. A CEO plays a pivotal role in shaping organizational culture. We often describe culture as "how things really get done when no one is looking." A CEO can proclaim values like agility, boldness, or collaboration, but if the people in key positions aren't naturally inclined or properly incentivized to act that way, those cultural aspirations will fall flat. By understanding the innovation profiles of their top team (and perhaps a layer or two below), a CEO can identify where they might need to reinforce or change the culture.

For example, imagine a CEO who has set a strategy for the company to become more customer-centric and innovative, moving away from a legacy mindset. However, through ISPI™ data, she discovers that her leadership team is overwhelmingly weighted toward Execution strength and very low on Ideation strength. In other words, her team's comfort zone is executing well-defined plans, not dreaming up new ideas. That insight explains why few groundbreaking ideas are bubbling up, it's not that her team lacks talent or work ethic; it's that collectively their style is to keep the trains running on time rather than question where the tracks should go next.

Armed with this knowledge, she can take concrete steps. She might establish a new "Pioneers Council," a rotating forum of high-ideation employees from various levels and departments (identified via ISPI™) who meet with her regularly to pitch outside-the-box ideas directly. She could also reshuffle roles or adjust her org chart to give some of these pioneer-types more visible platforms to influence product direction, rather than burying them in roles where they're only executing on others' ideas. By doing this, she sends a powerful signal to the whole company: We value visionary thinking at the highest levels. Over time, that encourages more people to speak up with ideas and reinforces that innovation is truly welcome, not just a buzzword on a poster in the break room.

Second, team composition. One of a CEO's most crucial jobs is getting the right people in the right seats on the bus. Yet many CEOs unintentionally build one-dimensional teams. They might hire and promote in their own image, or default to candidates with impressive credentials without considering how they complement or duplicate the existing team. Incorporating an innovation lens can prevent that.

Suppose a CEO is personally a bold, risk-tolerant visionary, in ISPI™ terms, a Pioneer with a high-risk orientation. That's a great trait for a chief executive, but if everyone on the top team shares that profile, you have a potential issue: who's minding the details and making sure the wheels don't come off? In such a case, that CEO might deliberately seek a COO or CFO who has a more cautious, detail-oriented profile to introduce balance. Conversely, consider a CEO who knows she's more of an Adapter, she prefers refining existing successes over chasing radical change and tends to be risk-averse. To avoid letting the company get too comfortable, she might surround herself with one or two senior leaders who are unabashed innovators, giving them license to challenge the status quo.

I recall a tech CEO who openly acknowledged that he was naturally conservative. He went out of his way to hire a flamboyant, big-idea Chief Strategy Officer and explicitly gave that person leeway to propose "crazy ideas." The CEO's role then became tempering and channeling those ideas into feasible plans. It was a productive tension they established on purpose. Had he not done that, one of two things could have happened: either he would have hired someone too similar to himself (resulting in a cautious leadership team missing opportunities), or he might have hired an innovator but grown frustrated or fearful of their ideas, leading to conflict or the innovator's departure.

By explicitly acknowledging their different styles up front and using an assessment to facilitate a frank conversation, this CEO and CSO built mutual trust and learned to leverage each other's strengths effectively.

Third, project oversight. The CEO is often the ultimate portfolio manager of major innovation initiatives. Even with dedicated innovation teams or a Chief Innovation Officer, big decisions, which moonshot projects to fund, which to pause or kill, how to allocate resources, often land on the CEO's desk. Understanding the human element adds a crucial dimension to these calls. If Project X is mission-critical and has stumbled in the past, the CEO should ask, "Who's leading Project X, and do they have the right innovation DNA for this kind of challenge? If this project requires breakthrough, out-of-the-box thinking, did we assign a breakthrough thinker to lead it, or just the person with the longest tenure in that department by default?"

I've seen instances where a strategic project was floundering. The CEO's intervention was not to cut the budget or change the goal, but to change the leader. In one case, a very methodical project manager was replaced with a more visionary leader who could inspire the team differently. In another, a brilliant but chaotic visionary was replaced with a structured executor who could get the project back on schedule. In each scenario, those decisions were guided by an understanding of the individuals' innovation styles, not just their titles or political clout. And importantly, the CEO had the authority to make that tough call and then protect the newly restructured team while it adjusted.

When employees observe that the CEO is actively and thoughtfully crafting teams for success, for instance, pairing complementary executives and even talking about the need for diversity of thought in town halls or staff meetings, it has an electric effect on the organization. People realize that innovation isn't just an R&D directive; it's woven into how the company operates. They also see that top leadership cares about how results are achieved, not just that they are achieved. That boosts morale and encourages the whole management chain to pay attention to team dynamics, not just KPIs.

Finally, a CEO who embraces this human-first, data-informed approach sets a precedent for the rest of the leadership. It shows that while gut instinct and experience remain valuable, they are being augmented by evidence and science when it comes to understanding people. That encourages other leaders to follow suit. Instead of the old way, "Jim and Karen aren't getting along, let's just separate them", managers might start to say, "Perhaps Jim and Karen have a style conflict. Let's dig into what that is and help them work through it, because their different perspectives could be an asset if we manage it right." This kind of analytical yet empathetic mindset trickles down from the CEO's example.

In summary, when a CEO fully buys into and uses innovation DNA engineering, they turn lofty strategic intentions into tangible actions. They ensure that the who and the how of execution are fully aligned with the what. It can mean the difference between a grand strategy that lives only on PowerPoint slides and one that truly comes to life in the organization.

The CHRO and People Leaders: A Data-Driven Talent Strategy for Innovation

If the CEO is the champion of this approach, the Chief Human Resources Officer (CHRO) and their team are the architects and enablers. In many organizations, HR has evolved from a back-office function into a true strategic partner. Nowhere is that more evident than in shaping a workforce that can meet future challenges. For a CHRO (or Chief People Officer, Chief Talent Officer, whatever the title), ISPI™ offers a powerful tool to weave into every stage of the talent lifecycle, from hiring and team formation to leadership development and succession planning.

One of the CHRO's primary mandates is to ensure the organization has the capabilities and culture needed to execute the business strategy. Traditionally, HR might approach that by defining desired competencies (e.g., "strategic thinking" or "collaboration") and then rolling out training programs to instill them. Those are still important, but ISPI™ provides a far more quantitative and nuanced way to assess and cultivate those capabilities.

For example, a CHRO could partner with an innovation consulting team (like ours) to map out the innovation DNA profiles of various employee groups, perhaps comparing high-performing project teams to average ones, or looking at why some innovation initiatives succeeded while others struggled. They might discover patterns: say, teams that consistently delivered breakthrough results all had at least one strong risk-taker paired with one strong risk-mitigator, whereas teams that failed were uniformly one or the other (all wild dreamers with no pragmatists, or vice versa). Armed with that insight, HR could create new guidelines for assembling teams. When forming a new product task force or innovation project, managers would be encouraged (or required) to include a balance of key profiles. This can be baked right into project planning checklists or leadership training, much like how HR departments institutionalized the idea of cross-functional diversity on teams, you now institutionalize cognitive and innovation-style diversity as a must-have for any high-stakes initiative.

In recruitment, a CHRO might carefully integrate innovation profiling into the hiring process. The emphasis is on carefully, you don't want to turn candidates off with an overly test-heavy process, or misuse data in a way that filters out good people. But for senior roles or positions where innovation is central, having candidates complete an ISPI™

assessment as one input can be extremely valuable. The key is how you use the data. A best practice is for HR to use the ISPI™ results to brief the hiring manager on a candidate's likely strengths and potential gaps relative to the existing team, not to make yes/no decisions in a vacuum. The conversation might be, "This candidate's profile suggests she thrives in ambiguity and moves fast (great for the role, since the team needs that), but she may be impatient with lengthy consensus-building. If we hire her, let's be mindful to onboard her accordingly." It's never about labeling someone as "bad at innovation" or disqualifying a candidate based on a profile. Every profile has strengths; the goal is to ensure that if you're hiring a leader into, say, a very chaotic, change-heavy environment, you either pick someone who shows they can thrive in that chaos or you prepare to support them appropriately if they don't.

I've seen CHROs use assessments in onboarding to wonderful effect. They'll sit down with a new executive and say, "Here's what makes you tick (according to this data), and here's what might frustrate or challenge you in our culture. Let's talk about both upfront." This kind of frank conversation can preempt months of misunderstandings. ISPI™ can make it even more targeted. For example, an incoming leader's report might indicate, "Prefers quick action over consensus." HR can then advise, "In our environment, some teams move methodically. We need your bias for action to speed them up, but be aware that initially you might come across as impatient. Let's plan some ways for you to introduce your style constructively." That's proactive onboarding at its finest, turning a potential culture clash into a productive integration strategy.

The CHRO is also the guardian of leadership development and succession planning. Here, the innovation DNA lens can be transformative. Many companies' succession plans are essentially lists of who might replace whom, largely based on performance in current roles and maybe a general leadership potential rating. By overlaying ISPI™

data, HR can ensure that tomorrow's leadership bench has the right mix of innovation traits for where the company wants to go.

Imagine the company's five-year strategy calls for doubling revenue through new digital products and global expansion. The next generation of leaders will need to be more globally savvy, digitally fluent, and innovative than the last. If HR reviews its pool of high-potential leaders through the ISPI™ lens and finds they're almost all cut from the same cloth, maybe they're all superb operators but not particularly visionary, that's a signal to take action. HR might start recruiting some different profiles into the leadership pipeline, or deliberately develop current high-pos to stretch their innovation skills. Perhaps send some of those operational stars to an innovation bootcamp, or rotate them through a stint in R&D or a startup incubator to broaden their perspective. Conversely, HR might identify a few hidden "Pioneers" deeper in the organization, talented people who haven't yet been noticed because they don't fit the traditional mold, and fast-track them into bigger roles to inject fresh thinking into the leadership pipeline.

Another area where ISPI™ aids HR is in employee engagement and retention. Research and common sense both tell us that people are more engaged when they feel they can use their strengths and contribute meaningfully. Consider an employee who's a natural Innovator (loves generating ideas and taking risks) but is stuck in a routine maintenance job. They're likely to disengage or leave out of frustration. On the flip side, think of a process-oriented executor who's suddenly thrown into a chaotic, undefined role, they may burn out or underperform. By working with managers to be mindful of their team members' innovation styles, HR can often suggest small tweaks that yield big improvements. Maybe that creative salesperson itching for novelty gets an opportunity to design a new sales campaign, tapping into her ideation strength. Maybe that systematic engineer who craves order is put in charge of optimizing the team's workflow, leveraging his knack for process improvement. In many cases, people blossom when given the right assignments, and the company benefits from higher innovation and efficiency. Plus, a culture that acknowledges and leverages different innovation styles is inherently more inclusive. It sends a message: We need all types of thinkers here, and your way of innovating is valuable. That sense of being valued is fundamental to retaining top talent.

On the flip side, HR can also use these insights to address performance issues or team conflicts. Instead of writing someone off as a "poor performer," it might turn out their innovation style is just a poor fit for their current team's dynamics. We encountered a case of a highly creative marketing manager who was labeled "not a team player" because she kept deviating from agreed-upon plans. She wasn't trying to be difficult; her ISPI™ profile revealed she was an extreme Innovator who just couldn't resist exploring new possibilities, even after the decision phase was supposed to be over. Once we recognized that, HR intervened. They worked with her and her supervisor to find a better outlet for her creativity (involving her earlier in the strategy phase of projects, where her ideation was an asset) and coached her on understanding when to stick to the plan during execution. Meanwhile, the team learned to appreciate her ideas at the right stage. The tension eased, and she went from being a flight risk to being seen as an "intrapreneur" who added real value. Without that reframing, the company might have lost a talented person, chalking it up to "she just didn't fit."

For a CHRO, championing the use of ISPI™ and innovation people analytics is an opportunity to elevate the HR function into a direct driver of business outcomes. It moves HR beyond administering engagement surveys and generic training programs, to actively shaping innovation success rates, the speed of project execution, and the organization's adaptability. In essence, it lets HR answer the CEO's question, "Do we have the people and culture to win?" with a confident "Yes, and here's the data, insights, and plan to back it up."

CIOs, CTOs, and Tech Leaders: Bridging Human and Technological Innovation

On the surface, one might assume Chief Information Officers (CIOs), Chief Technology Officers (CTOs), and other technology leaders are focused on systems, software, and infrastructure, the bits and bytes of innovation. But any seasoned CIO or CTO will tell you that the toughest part of introducing new technology is never the tech itself. It's the people and the processes around it. How many times have we seen a brilliant new system fail because employees wouldn't adopt it? Or a promising AI project get stuck in "pilot purgatory" because the organization wasn't culturally or talent-wise prepared to embrace it?

For technology leaders, aligning the human element with technological innovation is not a "soft" concern; it's a critical success factor for their initiatives.

One major challenge tech leaders face is change management. Let's say a CIO is rolling out a new enterprise AI platform intended to revolutionize how work gets done. The technical build might be flawless, but the rollout will succeed only if managers and employees actually embrace the new way of working. By collaborating with HR and leveraging tools like ISPI™, the CIO can identify who in the organization are natural early adopters of change, those whose profiles show high comfort with ambiguity and enthusiasm for new ideas. These individuals can be enlisted as change champions or pilot users to build momentum. Likewise, ISPI™ can spotlight who is more cautious and methodical, people who might resist or struggle with the change. Instead of labeling them "laggards," the CIO (with HR's help) can develop targeted support such as extra training sessions, pairing them with a tech-savvy colleague, or highlighting quick wins that speak to their perspective. For example, emphasizing how the new system reduces error rates and stabilizes processes appeals to someone who values accuracy and reliability. This approach is far more empathetic and effective than issuing a blanket mandate or delivering a one-size-fits-all town hall. It acknowledges that deploying technology is as much a human project as a technical one.

Furthermore, many CIOs and CTOs lead dedicated innovation teams or "labs" within IT, especially for tech-driven innovation. Using ISPI™ to staff these teams can prevent some classic pitfalls. A complex tech project often requires deep specialists, but if you fill a room only with like-minded engineers, you might end up with a technically elegant solution that nobody uses. Perhaps the team didn't consider user behavior, or they struggled to pivot when requirements changed. As a tech leader, you want to ensure your teams include a mix of innovation profiles. Some team members should represent the voice of the user, people high in collaboration and empathy, who can constantly relate the project to user needs. Others should bring a disruptor's mindset to challenge assumptions and avoid tunnel vision. And yes, you still need the disciplined project managers to keep the trains running on time. Striking this balance makes it much more likely that your innovations will be both cutting-edge and practical.

I remember a data analytics leader who realized he'd made a mistake in team composition. He had staffed an experimental analytics project with five people who were eerily similar: all detail-oriented, somewhat risk-averse folks who were brilliant with numbers but not inclined to talk about their work across departments. They produced an excellent model… but nobody paid attention to it, because the team hadn't socialized their findings or won buy-in. After that experience, he reshaped the team. He added a business analyst with a more extroverted, Pioneer profile to liaise with stakeholders and drum up interest, and brought in a creative data scientist who loved challenging assumptions to ensure the team explored fresh approaches instead of just refining the same model. The change in team DNA led to a change in outcomes: the model improved in unexpected ways (thanks to the creative input) and, critically, people in the organization heard about it and started using it (thanks to the outreach and enthusiasm of the new team members). The key was recognizing the initial one-dimensional team makeup and intentionally correcting it.

CIOs and CTOs also frequently straddle two worlds: operational stability and innovation. They're accountable for "keeping the lights on" (ensuring systems are up, secure, and compliant) and for pushing the company forward technologically. It can feel like a Jekyll-and-Hyde mandate. Many IT organizations handle this by having separate "run" teams and "change" teams, one focused on maintenance and reliability, the other on new development and transformation. ISPI™ can inform how you staff and manage those parallel streams.

For example, your "run" teams (think network operations, maintenance, cybersecurity monitoring) may benefit from more stabilizing profiles: Adapters who excel at refining and maintaining, risk managers who proactively prevent problems, execution-focused individuals who take pride in consistency and uptime. Your "change" teams (like a digital innovation group or an AI implementation squad) often need more Innovators who challenge the status quo, risk-takers who aren't afraid to experiment, and collaborative brainstormers who iterate quickly.

Of course, those two groups still need to work together, you don't want a wall between them with each side thinking the other "doesn't get it." This is where a people-centric approach helps bridge the divide famously addressed by DevOps culture. You can facilitate sessions where each side learns to appreciate the other's value. Show the operations folks that the innovators aren't just cowboys; they play a vital role in pushing the company into the future. Show the innovators that the operators aren't sticks-in-the-mud; they protect the business by ensuring quality, security, and reliability. I've seen IT departments where this kind of mutual understanding, fostered by a neutral framework like ISPI™, turned a combative relationship (development vs. operations throwing issues over the fence) into a collaborative one. When dev and ops actually respect each other's styles, changes roll out faster and more smoothly, because one side isn't constantly having to redo or battle what the other side did.

On a personal leadership note, understanding the innovation DNA of their teams can help tech leaders dispel stereotypes. CIOs/CTOs often hear complaints that IT folks "lack soft skills" or "don't speak business language." In truth, many tech professionals simply have a different communication and problem-solving style. If a CTO knows that many of her senior IT managers are, say, introverted soloist types who shy away from the spotlight (quite common among top technical experts), she can adjust expectations and communication strategies. She doesn't need to force those brilliant engineers to become extroverted storytellers overnight. Instead, she could pair each of them with a more outgoing business liaison who can help present their work to non-technical audiences. That way the tech expert can focus on deep analysis (their strength) without being pushed into an uncomfortable role, and the business gets the message through someone who naturally speaks their language. Often, IT staff are just told, "Be more business-oriented," which is vague and frustrating. A data-driven discussion about profiles can clarify how to bridge the gap: perhaps the solution is pairing complementary people or adding a new process, rather than trying to change someone's personality.

In summary, technology leaders succeed when technology is fully adopted and effectively leveraged by people. By using innovation DNA insights, CIOs and CTOs effectively become bridge-builders. They bridge current needs with future possibilities, technology initiatives with human readiness, and visionary ideas with execution know-how. They make sure the brilliant code and systems their teams develop actually translate into real business value, because the team behind the tech is engineered for success, and the users in front of the tech are guided and supported to embrace it. It's the ultimate combination of high-tech and high-touch leadership.

Strategy Executives: Driving Execution Through People Alignment

Many organizations have roles like Chief Strategy Officer (CSO), EVP of Strategy, or a strategy committee comprising top executives who drive major initiatives across the company, new market entries, product launches, mergers and acquisitions, digital transformations, and so on. For these strategy leaders, success isn't just about coming up with a brilliant plan; it's about orchestrating complex, cross-functional execution to make that plan a reality. And if you've ever led a big, cross-department project, you know the human element is frequently the hardest part. Different departments bring different micro-cultures, strong-willed leaders can clash, and teams often resist directives that come from outside their silo.

Strategy executives can use innovation DNA principles as a playbook for assembling and steering the "tiger teams" or task forces that drive strategic projects. Let's say the company's strategic priority is to launch a new digital service and enter a fresh market segment. A cross-functional team is formed with people from product, marketing, IT, operations, perhaps an external consultant or two. Typically, these individuals might be chosen simply because of their roles or availability, with little thought to how they'll work together.

Now imagine instead that you factor in ISPI™ profiles when building that team. Perhaps you notice that everyone initially selected comes from a similar background and tends to think alike (not ideal for generating fresh ideas). So, you swap in someone from a different geography or business unit who has a more innovative profile to shake up the perspective. Or you realize the team is stacked with big-picture visionary types but is light on detail-oriented executors. So, you add a project manager known for strong follow-through to keep the group on track and translate ideas into action. Even one or two intentional adjustments like these can make the difference between a team that flounders and one that flourishes.

Once the team is underway, strategy leaders can use the vocabulary of innovation styles to mediate conflicts and keep things moving. High-stakes projects often have tense moments, say the finance lead keeps shooting down the marketing lead's bold ideas as too risky, causing frustration. If you, as the strategy head, know from their ISPI™ profiles that this dynamic was predictable (the finance person is a risk-managing Adapter and the marketing person is a risk-taking Innovator), you can step in and reframe the discussion in a way that values both contributions. You might say, "We absolutely need your caution on the finance side, that's protecting us from reckless moves. And we need your boldness on the marketing side, otherwise we might only make tiny improvements. How about we design a small pilot experiment that gives us some data? It's a controlled risk that might satisfy both viewpoints." By effectively translating between their styles, you prevent a personal clash from derailing the project.

So many strategic initiatives stumble not because the strategy was flawed, but because team members lock horns and then either compromise to a watered-down middle or start avoiding each other entirely. A leader who understands the team's innovation DNA can coach everyone through these friction points. Productive tension between differing perspectives, when managed well, creates better outcomes than one-sided thinking ever could. The key is to harness that tension and keep it creative, not let it turn destructive.

Another benefit for strategy executives is in project risk assessment. Typically, when evaluating project risk, we consider technical risks, market risks, financial risks, and so on. We suggest adding human configuration risk to that checklist. In reviewing a portfolio of initiatives, ask: Does each important project have a balanced human engine behind it? For example, if a critical project team is drawn entirely from one department, do they have enough diversity of thought? If a key initiative is being co-led by two strong-willed executives who have very similar profiles, perhaps both are highly controlling and low on collaboration, is there a risk they won't gel or will create a bottleneck?

With those insights, a strategy leader can preempt problems. I once advised on a large transformation program that initially had two co-leaders: one from operations and one from technology, a common setup to ensure buy-in from both sides. On paper, it made sense. But our assessment suggested a looming issue: both leaders turned out to be ultra-driven, command-and-control personalities. Each was used to being the sole captain, and neither was particularly open to feedback. We foresaw either a power struggle or at least a lack of coordination because they'd be pulling the project in slightly different directions.

Armed with this information, the executive committee made a smart adjustment. They appointed one of those individuals as the single leader of the program and reassigned the other to a different, critical role that still leveraged his strengths but removed the day-to-day friction of having two heads. They also introduced a "project integrator" with a high-collaboration profile to act as connective tissue, making sure all workstreams stayed aligned and information flowed. The transformation progressed far more smoothly than similar efforts in the past, and importantly, the two senior leaders remained on good terms (avoiding a scenario where the company might have had to eventually pick between them).

This kind of proactive human risk mitigation is exactly what strategy leaders are well positioned to do. They have the bird's-eye view across silos and the clout to reshuffle team assignments if needed.

In essence, for strategy executives, incorporating innovation DNA engineering is about stacking the deck in favor of execution. It's a systematic way to reduce the execution gap that plagues so many strategies. It also sends a message to the whole organization: strategy isn't being cooked up in an ivory tower, detached from reality. It's being executed by real people with particular strengths and weaknesses, and those realities are going to be acknowledged and managed rather than ignored. That makes the strategy process feel more human and achievable, rallying teams around a common goal rather than intimidating them with edicts from on high.

Rising Leaders and Mid-Level Managers: Accelerating Your Path to the C-Suite

Not everyone reading this is a CEO or a C-suite member. Many leaders sit in the middle of organizations, managing teams and large parts of the operation, while aspiring to higher roles. If that's you, perhaps a Director, Senior Director, VP, or similar, you might be thinking, "This all sounds great, but how can I influence it from where I sit?" The encouraging reality is that you can adopt many of these practices within your own sphere and, in doing so, distinguish yourself as a forward-thinking leader. That is exactly the kind of mindset upper leadership values and promotes.

First, by using ISPI™ insights with your own team, even informally, you can boost your team's performance, which reflects well on you. For example, you might not have the budget or authority to formally assess everyone, but you can start by observing and mapping your team's apparent innovation styles using the concepts we've discussed.

You likely already have a sense of it: John is the ideas guy who hates details. Maria is the project manager who keeps everyone on schedule. Alex is the skeptic who calls out risks. Priya is the peacekeeper who makes sure the team collaborates.

Once you have that rough map, validate it through conversation. Ask your team members whether they see these as their strengths and how they prefer to work. You might be surprised. Sometimes a team member will say, "Actually, I love brainstorming, but I've never spoken up because this team already has strong voices." Suddenly, you've uncovered an underused Pioneer.

By openly discussing these preferences, you create a new kind of team dialogue. You can say, "We have a big challenge this quarter. Let's intentionally leverage what each of us does best. And if we're missing something, like if none of us are naturally detail-oriented, let's acknowledge that and double-check our work, or bring in someone from another team to review our plan." This level of self-aware management is rare, but when it happens, teams feel the difference. There are fewer dropped balls, less frustration, and more mutual respect.

And who gets the credit for that? You, the leader who orchestrated it. Your superiors will notice not only the improved results, but also that your people are engaged and growing. That combination is a strong signal of leadership readiness for bigger roles.

Second, knowing your own ISPI™ profile (or at least having a working theory about it) is a huge asset in career development. Perhaps you've discovered that you're a Builder who excels in execution but hasn't had many opportunities to flex your creative muscles. You could proactively seek assignments that pair you with a visionary leader, allowing you to both support and learn from that style. You might also work on developing some of your weaker areas, not to change who you are, but to become more versatile.

For instance, if you know ambiguity makes you uncomfortable, you could volunteer for a slightly ambiguous project as a growth opportunity, signaling to senior leaders that you're intentionally broadening your range. You can also communicate to your boss or mentor the conditions that help you do your best work. For example, "When I get to brainstorm with others first, I tend to come up with stronger strategies. But if I have to implement without that ideation phase, I might not challenge assumptions enough. So I'm building a quick pre-mortem brainstorm into my project process."

That kind of self-awareness is gold in performance discussions. It moves you from being a passive recipient of feedback ("you need to be more strategic") to an active co-owner of your development ("I've identified that I'm highly execution-focused. I'm incorporating more strategic thinking into my process, and here's how"). Trust me, bosses love that. It signals maturity and real leadership potential.

Mid-level managers are also the crucial translators of top-level vision into ground-level reality. If you're attuned to innovation DNA, you can better interpret and rally your team around initiatives coming from above. Say the CEO launches a transformation program and you sense anxiety on your team, people mumbling, "We've seen these come and go." As a savvy leader, you might recognize that this anxiety is coming from the more risk-averse, stability-oriented members of your team.

You can address that directly by acknowledging their style. For example: "I know a lot is changing, and that can be unsettling. It's actually a strength that we have people thinking about the stability of our operations. Let's surface your concerns and see how we can mitigate risks while still moving forward." At the same time, you likely have a few change enthusiasts who are eager to jump in. You can channel their energy by positioning them as change ambassadors, keeping them engaged and using their optimism to balance the team's overall mood.

This is emotional intelligence combined with innovation intelligence. Managing that mix well is what separates leaders who can implement strategy from those who only talk about it. As you build that reputation, you'll find more doors opening. It becomes, "Give the new AI pilot to Alex's team. They always seem to find a way to get things done."

Additionally, your familiarity with these concepts can help you interface with senior leadership more effectively. When you can articulate your team's needs or achievements using this language, it resonates. Imagine being in a meeting with executives and saying, "Our division has a lot of great Builders. We execute reliably. To hit next year's targets, we're adding more exploratory projects, and I've identified a couple of creative thinkers in our ranks to lead those. We're also partnering with R&D to bring in fresh perspectives. We're intentionally balancing our innovation portfolio."

That kind of statement does two things. It signals that you understand how work gets done, not just what needs to be done, and it subtly demonstrates that you're already practicing this discipline of engineering innovation into teams. Essentially, you're speaking the language of the leadership book they're reading (maybe even this one), and showing that you're ahead of the curve. This can accelerate your path to promotion or to being tapped for special assignments.

In short, mid-level leaders who embrace innovation DNA principles become multipliers of innovation within the organization. They turn their teams into high-performing units and create success stories that senior leaders want to replicate. In doing so, they naturally rise to greater responsibility because the organization sees them as contributors to the culture and performance everyone aspires to build. It's a classic case of leading by example. By engineering your own team's success, you often find yourself asked to do the same at a larger scale, effectively auditioning for that next-level role.

Every Employee: Empowering Personal Innovation and Growth

We've talked a lot about leaders, but let's look at how this affects every employee, at any level. Not everyone will take an ISPI™ assessment or be part of a formal team redesign. Companies roll these programs out gradually. But the philosophy behind them is inclusive. Everyone has innovation strengths that can be harnessed, and everyone benefits when work is aligned with their natural capacities. So even if you're not a manager or not in an "innovation" role, understanding these concepts can help you navigate your career and contribute more meaningfully.

For an individual employee, learning about your innovation DNA is like getting a personalized user manual for your work style. Have you ever been frustrated at work and not entirely sure why? Perhaps you're a creative soul feeling stifled by bureaucracy, or a meticulous planner drowning in a chaotic project. It's easy to internalize that as a personal failing ("I'm bad at this job"), or to blame others ("My boss is disorganized"). But when you realize, "My strengths just aren't being used here. How can I change the situation?" you move from victim to problem-solver.

Maybe you raise the issue with your supervisor and offer to take on a different set of tasks that better fit your strengths. Or you team up with a colleague to tackle work together, with each of you handling the aspects you're best at. For example, if you know you're not great at structuring a large task but excel at generating options and ideas, partner with someone who enjoys creating project plans. Offer to help them brainstorm if they help you with structuring. It's a fair trade that benefits both sides.

When people begin doing this kind of strengths-trading at a grass-roots level, the entire organization becomes more effective and, frankly, more humane. Work becomes less about hiding weaknesses and more about trading strengths.

Knowing your innovation preferences also helps you seek out the right opportunities. In a large organization, there are often internal openings or special projects. If you're someone who thrives on novelty and risk, you might volunteer for a pilot project involving a new technology. If you prefer steady improvement, you might gravitate toward a continuous improvement task force. One path isn't valued over the other, they're simply different, and both are needed.

Without this awareness, people often feel pressured to pursue the "hot," high-profile project even if it doesn't suit them. Others avoid new opportunities altogether due to self-doubt, missing a chance to shine in their natural element. With an innovation DNA lens, you can make more informed choices.

I recall a young analyst who was offered two assignments. One was on a fast-paced, creative marketing campaign team that involved lots of brainstorming. The other was on a process optimization project that was methodical and data-driven. He was torn because the marketing role sounded exciting and highly visible, but something made him hesitate. After learning more about his style through a workshop, where he discovered he leaned toward the analytical, structure-loving end of the spectrum, he chose the process project and excelled. He was promoted shortly afterward for the improvements he helped implement. Had he chosen the marketing assignment, he might have been unhappy or overshadowed by more naturally creative colleagues. That bit of self-knowledge saved him, and the company, from a misstep.

For employees, embracing this approach also means feeling valued for who you are. In many workplaces, certain types of people seem to get all the glory, maybe the extroverted creative types in some companies, or the hard-driving number-crunchers in others. That can leave those of a different mold feeling second-class. A culture that actively talks about and uses diverse innovation strengths sends a very different message. Your way of thinking is a strength, and here's how it fits into the big picture.

The quiet, detail-focused person who rarely speaks up in meetings might suddenly find colleagues seeking her out and saying, "We've done a rough brainstorm, but we need your critical eye to poke holes in it and make it better." How do you think that makes her feel? Empowered. She's needed and included precisely because of who she is. Meanwhile, the big talkers learn that they need those quieter analysts to succeed, which fosters mutual respect across the team.

This perspective also supports personal growth and learning. An employee who understands their profile can chart a smarter development path. If they aspire to be a manager and know, for example, that they prefer solo work, they can deliberately build their collaboration skills over time. The goal isn't to become a social butterfly, but to be effective in group settings when required. They might partner with a mentor known for strong team leadership and say, "I'm trying to get better at facilitating brainstorming sessions, which isn't my strong suit. Can I learn from you or even sit in on one of your sessions?"

That kind of targeted development is far more effective than generic advice like "improve your communication skills." It's specific and intentional. This is where I want to grow. And because the employee starts from a clear understanding of themselves, they can measure progress in ways that matter to them. Maybe after a few months, those sessions feel less daunting and even a bit enjoyable. That, by any measure, is a win.

Finally, when every employee has some grasp of these principles, workplace interactions improve. People begin to attribute differences to style rather than malice or incompetence. Instead of thinking, "Bob is always slowing us down with all his questions, ugh," it becomes, "Bob has a cautious style, which can actually save us from mistakes. Let's address his questions, and maybe our plan will improve." Or instead of, "Alice keeps proposing crazy ideas just to show off," it's, "Alice is a big visionary thinker. Even if not every idea is feasible, one of them might be a game-changer, so let's hear her out and then weigh them."

This shift in perspective reduces interpersonal friction and leads to more constructive outcomes. In a sense, it democratizes innovation. Everyone sees their role in it, not just the R&D lab or the strategy department. And when innovation is democratized, an organization truly becomes agile and resilient. Ideas and improvements can come from anywhere, execution responsibility is shared broadly, and adapting to change becomes part of the fabric of daily work rather than a special emergency mode.

In sum, whether you're a new hire or a veteran employee far from the C-suite, understanding and applying innovation DNA concepts can make your work life richer and more effective. It's about becoming the best version of your innovative self and finding the place where that self can do great things. Collectively, when everyone is doing that, it propels the organization to a higher level of performance and engagement.

From the Engine Room to the Bridge: A Unified Approach

Bringing it all together, the message of this chapter is that engineering human innovation capacity is a unifying strategy. It bridges the traditional gaps between the top floor and the shop floor, between strategy and execution, between HR and the business, and between technology and people. When a company adopts this approach, you can feel it at every level.

The Board talks about talent and innovation in the same breath as financial results. The CEO becomes not just the chief strategist but the chief builder of teams, consciously shaping culture and leadership to meet the future. The CHRO and people leaders deploy data and insight to fine-tune the organization's most complex system, the human system. Tech leaders ensure the brilliance of their teams is aligned with the people ready to use it. Strategy leaders de-risk big bets by making sure the right humans are on the job. Mid-level managers turn their units into innovation engines and become tomorrow's senior leaders. And employees at all levels see a workplace where their unique contributions are essential to collective success, motivating them to engage and grow.

This is a departure from business as usual. It's a new discipline at the intersection of innovation science, human-systems engineering, and execution strategy. It doesn't fit neatly into conventional buckets. It's not just a training program, not just a re-org, and not just a tech tool. It's all of those woven together. And that is precisely what makes it powerful.

Companies that embrace this holistic, people-first approach find that things begin to click. Projects that once stumbled now hit their stride. Time-to-market improves. In some cases, we've seen complex project timelines cut by 50, 60%, not because people work longer hours, but because they work smarter together, with less friction and more intention. Decision cycles speed up because the right voices are in the room from the start, with less second-guessing later. Importantly, morale often climbs. People feel successful and seen, creating a positive feedback loop. Success breeds confidence, which breeds more innovation.

A real-world litmus test is how organizations fared during sudden crises or unexpected opportunities. Those that had this kind of human-centered agility weathered events like the COVID-19 pandemic or supply chain disruptions more smoothly. They could reconfigure teams, adapt roles, and innovate on the fly because they knew their people well and trusted them in the right roles. Others floundered, scrambling to figure out who could do what in a pinch.

In the coming AI-driven era, such agility will be even more crucial. Roles will shift as automation takes on certain tasks, and entirely new kinds of jobs will emerge. The companies that navigate this best will be those that understand the core DNA of their talent. If you know what makes someone innovative and adaptive, you can move them to where those traits are needed next, even if the job title changes.

Ultimately, making innovation DNA an everyone, everywhere imperative creates a culture that is Human First. AI Forward. which, not coincidentally, is the vision that inspired us to write this book and launch this movement. It means we don't lose sight of people amid the race for the next tech breakthrough. Instead, we put people at the forefront of driving technology in thoughtful ways. Every person is regarded as an innovator in their own right, and the role of leadership is to orchestrate these innovators in harmony. It's a vision of collective genius, engineered through understanding and trust, and unleashed through deliberate strategy.

As we wrap up Part II of this book, take a moment to reflect on your role in this picture. Whether you're a board member thinking about governance, a CEO pondering your next re-org, a manager eyeing a promotion, or an employee wondering how to make a bigger impact, there's a takeaway here for you. Innovating successfully in the AI era is not someone else's job. It's yours and mine. The good news is that we now have better tools and methods to do it, along with a map (or perhaps a periodic table) that lights the way forward.

Key Takeaways (Chapter 6):

- Board-Level Strategic Insight: Innovation capacity is a strategic asset that boards should monitor as closely as financial metrics and technological capabilities. By understanding the innovation DNA of leadership teams, boards can better foresee and mitigate strategic risks, such as a homogeneous leadership style that creates blind spots, and press for the right mix of talent to fulfill the company's long-term vision. When boards treat human innovation factors as a priority, they set a tone that permeates the organization.

- Executive Alignment and Diversity of Thought: CEOs and C-suites thrive when they engineer their teams with complementary strengths. High-velocity innovation at the top requires a balance of visionaries and operators, risk-takers and risk managers, creative thinkers and structured planners. Deliberately building that balance, and addressing any gaps, leads to faster, better decisions and a culture where bold ideas and prudent execution coexist. Leaders such as CHROs can integrate ISPI™ data into hiring, development, and succession planning to ensure the organization's human capital is aligned with its innovation goals, making talent strategy a direct enabler of business strategy.

- Technology and Strategy Through a Human Lens: CIOs, CTOs, and strategy executives often find that their initiatives succeed or fail largely because of people factors. By using innovation DNA insights, tech leaders can drive adoption of new systems by identifying change champions and tailoring support for different working styles. They can also compose development teams that truly deliver value by intentionally mixing creative and stabilizing forces. Strategy leaders, in turn, can de-risk major projects by assembling teams with the right chemistry, preempting conflicts, and maintaining momentum. In short, human configuration becomes a core part of project planning, greatly increasing the odds of successful strategic execution.

- Empowering Mid-Level Leaders: Mid-level managers and rising leaders can apply these principles within their own teams to create near-term wins and demonstrate leadership acumen. By aligning team members with roles that match their innovation strengths and openly addressing differences in working styles, they unlock higher performance and engagement. These engineered team successes not only drive results but also showcase a manager's ability to lead in the new way companies urgently need. It's a path for emerging leaders to accelerate their careers by becoming catalysts of an innovation culture within their departments.

- Every Employee as an Innovator: A human-first approach to innovation recognizes that every employee has a role in the company's innovation ecosystem. When people understand their own innovation profile, they can seek opportunities that suit their strengths, collaborate more effectively with colleagues of different styles, and actively contribute ideas and improvements. This leads to a more inclusive and energized culture. Quiet planners, bold creatives, careful evaluators, and enthusiastic experimenters all see a place for their talents. The result is an organization humming with innovation at all levels, not just in isolated pockets, and a workforce that feels valued and engaged. In the AI era, this broad-based human agility will be what separates the winners from the rest, as companies that harness everyone's strengths will innovate faster and adapt more intelligently to whatever the future brings.

"Innovation becomes predictable the moment
we stop betting on technology and start
engineering people."

— **Greg Brisco**

CHAPTER 7

❧

Innovation - Engineering a People-Driven Innovation Engine

Innovation isn't a lucky lightning strike of genius. It's something you can deliberately design and build. In an era where advanced AI tools sit on every desk, technology alone no longer gives you an edge. People do. The way you configure human talent and teamwork is what turns accessible tech into game-changing innovation. Human First. AI Forward. is more than a slogan, it's a strategy. In this first pillar, Innovation, we focus on engineering that strategy into your teams. The goal is to make innovation predictable by shifting the spotlight from shiny new tech to the people at the heart of it all.

Innovation as a Team Sport

The myth of the lone genius dies hard. We love the story of a brilliant individual conjuring the next big thing in a garage. But in reality, innovation is a team sport. Breakthroughs ignite when different strengths collide and complement each other. You need bold Pioneers who dream big and push into the unknown, and steady Builders who turn those dreams into reality brick by brick. You need audacious risk-takers to challenge the status quo, and careful risk-managers to ensure not every bet sinks the ship.

Now picture a team overstuffed with Pioneers. You'd get a hundred ideas a minute and zero follow-through. Flip it around: a team of nothing but Builders executes flawlessly on incremental improvements but produces no bold moves. Balance isn't just a feel-good concept; it's strategy. With data from the Innovation Strengths Preference Indicator (ISPI™), you can orchestrate that balance instead of leaving it to chance. It's innovation by design. When you deliberately mix visionaries with pragmatists and thrill-seekers with safety nets, you build a deeply human, endlessly adaptive engine that consistently generates great ideas and actually makes them happen.

The Innovation Continuum

Not all innovation is created equal. Different goals call for different approaches. I like to think of it in three flavors:

- Evolutionary: Small, steady improvements to what you already do (think of those yearly smartphone updates that get a little better each cycle).
- Expansionary: Stretching what you're good at into new markets, new products, or new applications.
- Revolutionary: Big, bold paradigm shifts that redefine an industry or create an entirely new one.

Each type demands a unique mix of innovation DNA. For an evolutionary update, lean on your Builders and risk managers, they'll polish and perfect, making sure nothing breaks. But sprinkle in a Pioneer or two to prevent groupthink and keep things fresh. If you're going for a revolutionary moonshot, flip that formula: stack the team with Pioneers and risk-takers who aren't bound by "how we've always done it," but don't forget to have a Builder in the room to ground that vision in reality. Expansionary efforts often need a 50/50 blend of creative minds and disciplined executors, because you're venturing outward while still relying on core strengths. The Innovation Continuum ensures you align your team's talent with the mission at hand, matching the people formula to your ambition.

Creative Tension as Fuel

Great innovation teams embrace a bit of tension, the productive kind. If everyone politely agrees all the time, you're probably not innovating at full throttle. The goal isn't chaos; it's constructive friction. That means deliberately pairing opposites. Sit your cautious risk mitigator next to an audacious risk-taker, and your visionary idea generator beside a meticulous planner. Yes, they will disagree, and that's the point. That friction can spark ideas neither would have created alone.

Of course, tension left unmanaged can blow up in your face. The key is to manage it with intent. Set ground rules and build mutual respect. Use data (like ISPI™ profiles) to anticipate where clashes might arise, then design your teamwork to channel those differences productively. When everyone sees contrasting styles as complementary assets rather than personal conflicts, creative tension becomes pure fuel. It pushes ideas further than they'd ever travel in an echo chamber of like-minded thinkers.

Predictable Innovation

Here's a bold statement: innovation can be made predictable. Not in the sense of guaranteeing exactly when the next billion-dollar idea will hit, but in reliably creating the conditions for innovation to thrive. How? By measuring and managing the human factors behind it.

Think about metrics you can track for a team:

- Innovation Velocity: How fast does an idea move from concept to prototype to market? If it takes too long, something's bogging your process down.

- Learning Rate: How quickly does the team absorb lessons from experiments and setbacks? If they keep repeating mistakes, they're not learning fast enough.
- Trust Level: Do people feel safe sharing wild ideas and admitting when something isn't working? If not, fear of failure is likely stifling creativity.

These are tangible indicators of an innovative culture. When one of these metrics lags, the root cause is almost always human. Maybe the team isn't aligned on the vision. Maybe fear of failure is choking initiative. Or maybe silos and turf wars are slowing collaboration to a crawl.

The best leaders treat every setback or slowdown as data, not defeat. If a project stalls, they ask "Why?" and look for human causes. Perhaps Team A and Team B aren't communicating, or everyone's waiting for permission to take a risk. Whatever the issue, they fix it. They create feedback loops so every misstep fuels the next success. In doing so, innovation stops being a happy accident and becomes part of the company's DNA, a steady rhythm of trial, learning, and triumph.

Case Example: Raytheon's Innovation Challenge

Even industry giants can stumble when their human systems fall out of tune. Take Raytheon, the global technology and defense powerhouse. They launched an internal AI innovation challenge and were flooded with hundreds of ideas from across the company. Sounds fantastic, right? But when the dust settled, only two projects got funded, and neither went anywhere. It was a classic innovation bottleneck, and the culprit wasn't lack of budget or technical know-how. It was people.

It turned out the committee selecting ideas was packed with risk-averse Builders. They reflexively shut down the bolder proposals, not because those ideas were bad, but because the committee's makeup skewed too safe. Meanwhile, the two project teams that did get funded were full of visionary Pioneers but had almost no Builders to execute the plans. Raytheon had inadvertently set itself up to fail: the gatekeepers were too cautious, and the doers were all dreamers.

Our team helped them flip the script, using ISPI™ data to guide the changes. First, we rebalanced the selection committee by bringing some adventurous Pioneers into the mix so bold ideas got a fair shot. Next, we re-engineered the project teams, ensuring each had a healthy blend of big thinkers and get-it-done executors. The results were night and day. Within two years, funded projects jumped from two to twelve (a 400% increase), and those projects delivered a nine-figure boost in revenue. The technology hadn't changed. The budget hadn't changed. We simply rewired the human system, and it unlocked an explosion of innovation.

The lesson is simple: when you humanize innovation by putting the right people in the right configuration, you transform creativity from a lucky strike into a repeatable, measurable engine for growth. But even with the perfect team and process, innovation won't stick unless your culture supports it. If taking smart risks gets you applauded instead of punished, and if failing fast is treated as learning fast, then innovation becomes a way of life. In that kind of environment, people aren't afraid of new AI tools, they're experimenting with them. They aren't hiding failures, they're sharing what they learned. That's a human-first culture that technology can amplify a thousandfold.

Key Takeaways (Chapter 7)

- Innovation is systematic, not serendipitous. You can deliberately engineer innovation by focusing on people. Build teams with a mix of Pioneers (idea generators), Builders (executors), and everything in between. Diversity of thought isn't a nice-to-have, it's the fuel for creative breakthroughs.

- Match the team to the mission. Use frameworks like the Innovation Continuum to align your talent with the project type. Evolutionary, expansionary, and revolutionary efforts each require a different game plan. Make sure your team's innovation DNA fits the goal at hand.

- Tension can be productive. Don't avoid disagreements, design for them. Pair up contrasting profiles so that creative friction becomes a spark instead of a stumbling block. Encourage respectful debate and let opposites challenge each other to reach better solutions.

- Measure the human factors. Track things like innovation velocity, learning rate, and trust levels. When something's off, it's usually a people issue. Treat setbacks as feedback: fix the underlying human factor, and innovation will follow.

- Reconfigure for results. Raytheon's 400% project boost wasn't magic, it was human engineering. Change the mix of people and you change the outcomes. The right human configuration turns AI and tech investments into exponential returns.

- Culture makes it stick. When people feel safe taking risks and sharing ideas, innovation stops being a one-time effort and becomes woven into your organization's DNA. Create a community where curiosity and collaboration are the norm, and watch innovation flourish.

"Change succeeds when people
are treated as co-creators,
not collateral."
— **Greg Brisco**

CHAPTER 8

❧

Leadership - Re-Architecting Leadership for the AI Era

The era of the lone hero CEO is over. For decades, organizations idolized the leader who single-handedly steered the ship, the visionary founder, the decisive chief at the top. In the AI era, that model simply can't keep up. Things move too fast and complexity runs too deep for any one person to have all the answers. Leadership can no longer rest on a single pair of shoulders. It has to function as a distributed, adaptive system, think of it as an organizational operating system for leadership that's as fast and flexible as the environment around it.

From Hero to System

Imagine your organization as a ship in stormy seas. If you rely on just one captain at the wheel while waves are crashing, you're in trouble. You need an entire crew of leaders on deck, each empowered to adjust the sails, steer clear of hazards, and keep the ship on course. In practical terms, decisions need to be made at all levels, in real time, by people you trust to lead in their own domain. Leading in the AI era means designing a system of many leaders, not elevating one hero.

The hard truth is that many big transformations fail because leadership hasn't caught up to this reality. Roughly 70% of digital transformations fail, and rarely because the tech didn't work, they fail because decision-making is too slow and too centralized. The pace of change is blistering, but at many companies the pace of leadership is stuck in molasses. AI initiatives demand rapid-fire decisions, constant iteration, and the courage to pivot on a dime. If your organization still waits on a single decision-maker for every move, you've created a massive velocity gap. In the time it takes a traditional leader to approve a plan, a more agile competitor has already tried it, learned from it, and moved on.

Nonlinear Leadership

At Humanize Innovation, we use a concept called "nonlinear leadership" to close that gap. Nonlinear leaders operate on two levels at once: they maintain a long-range vision and empower quick action on the ground. It's like playing chess by designing the game board rather than trying to move every piece yourself. As a nonlinear leader, you set the vision and define the guardrails, then let your people make decisions when the moment is right.

This kind of leadership means giving up the need to have your hands in every decision. It's about trust and clarity. Train your people, give them data and clear principles, then let them play. They will sometimes make moves you wouldn't, and that's okay. If you've done your job well, those moves will advance the strategy in ways one person alone could never manage. In the AI era, collective intelligence beats solitary brilliance.

And yes, it requires humility. One of the top barriers to AI success, as Gartner notes, is that many leaders simply don't understand AI deeply enough, and you can't wing it at the top anymore. The smartest executives I know are the first to say, "I don't know, let's bring in someone who does," or "You understand this area better than I do, so you decide." They listen more than they talk. They empower others to lead in areas where those people are the experts. This isn't weakness; it's wisdom. By harnessing the full brainpower of your organization instead of just your own, you make far better decisions, far faster.

Engineering Leadership Teams

Leadership isn't just about the CEO. It's about the chemistry of the entire executive team. I often compare the C-suite to a company's brain: if every part of that brain works the same way, you're going to miss a lot. Complex problems will blindside a homogeneous team because no one has a different perspective to catch them. Homogeneity is fragility. If everyone around the table thinks and acts alike, your leadership system will crack at the first big surprise. But when you have a diverse set of minds, different leadership styles and innovation DNA, at the helm, you build resilience right into your strategy.

Take Hallmark, the greeting-card and gifts giant. A few years back, Hallmark's board faced a tough succession dilemma. They had plenty of seasoned executives in line for the CEO role, but they worried that none of those usual suspects had what it would take to lead the company into a digital, AI-powered future. The internal candidates were all excellent operators, but not exactly bold innovators. Rather than force a square peg into a round hole, Hallmark's board tried something different: they used data to broaden the search. They ran ISPI™ assessments not just on the C-suite, but deeper in the organization, and discovered a mid-level manager who scored off the charts in visionary thinking, adaptability, and mentorship. On paper, this individual wasn't an obvious choice, he lacked the big-title experience others had. But his innovation DNA was exactly what Hallmark needed.

Recognizing this hidden talent, Hallmark's leadership nurtured him. They gave him high-profile projects, put him in charge of a major digital initiative, and paired him with executive mentors to round out his experience. Over the next few years, he rose into a senior role and eventually became COO. Under his influence, Hallmark launched new digital product lines and campaigns that rejuvenated the brand. That mid-level manager, whom most companies would have overlooked, became the transformational leader Hallmark was looking for.

The takeaway here is profound: diversity pays off at the top. Many leadership teams fall into the trap of groupthink, staffed with like-minded, risk-averse executives. Hallmark avoided that by injecting a different kind of thinker into their top ranks. By looking beyond the obvious candidates and evaluating people's innovation strengths, they found a game-changing leader hiding in plain sight. Often, the future leader your company needs is already in your midst, you just have to be willing to break the mold to find them.

Measuring Leadership as a System

We manage what we measure, and that goes for leadership too. If you're treating leadership as a system, you should put some gauges on it, essentially a dashboard for the health of your top team. A few metrics we use are:

- Decision Cycle Time: How long does it take for a proposal to turn into a decision and then into action? If it's taking weeks or months, you have a drag in your system that could be fatal in the AI age.
- Alignment Index: How in-sync is your leadership team on the big priorities and on your organization's appetite for risk? If half the team is charging ahead on a bold vision while the other half is slamming the brakes, that disconnect will tear your strategy apart. You need true alignment on the fundamentals (even if there's healthy debate on the details).

- Leadership Fragility Index™: What cracks might be forming in your leadership foundation? Are there signs of burnout at the top? Are trust issues or unresolved conflicts simmering beneath the surface? These fault lines, if ignored, could lead to an implosion at the worst possible time.

By tracking these kinds of metrics, CEOs and boards can tune the leadership system proactively. If Decision Cycle Time is lagging, dig in and find out why, maybe authority isn't as distributed as you thought. If the Alignment Index is low, perhaps your vision wasn't clearly communicated or genuinely bought into by everyone at the table. If Fragility is high, it's time for some frank conversations (or maybe a break for an overtaxed team) before things snap. The point is to make leadership performance visible and then manage it with the same rigor you'd apply to a financial budget or a product roadmap.

Leading People Through Complexity

All the systems and data in the world won't matter if your people aren't with you. Let's be real: AI can spook people. When employees hear that AI is coming for their workflows, some immediately think, "Am I about to be replaced by a robot?" Fear of change, fear of losing relevance, fear of the unknown, these are natural human reactions, and leaders ignore them at their peril. Leading in the AI era isn't just about tech and strategy; it's also about guiding your people through their anxieties and into new opportunities.

I saw this firsthand at a financial services firm that rolled out an AI-driven customer service platform. Almost immediately, whispers started in the halls: "AI is code for layoffs." Morale plunged. Sensing the fear, the CEO did something simple but powerful: he addressed it head-on. He stood in front of the entire company and promised that no one would lose their job because of AI. Instead, as automation took over the drudge work, every employee would be trained for new, higher-value roles. That pledge changed the atmosphere almost overnight. Skepticism turned to relief, then to excitement. People realized AI wasn't there to eliminate their jobs, it was there to elevate their jobs.

The lesson? Trust is your secret weapon. Leaders who communicate transparently and then back up their words with action (in this case, funding extensive upskilling and training) earn a level of trust that makes even the most daunting changes feel surmountable. When your team truly believes you have their back, they will run through walls for you. They'll embrace that new AI tool and master it, because they see it as a path to growth, not a threat.

Key Takeaways (Chapter 8)

- Leadership is a team sport, not a solo act. The age of the all-knowing, lone-wolf executive is over. The best organizations design leadership as a distributed system of empowered people, rather than a single point of failure at the top.
- Speed and adaptability are the new gold. If your decision-making can't keep up with the pace of change, you're in trouble. Nonlinear leadership, balancing big-picture vision with decentralized action, closes the gap between foresight and fast execution.
- Diversity equals resilience. A homogeneous leadership team will get blindsided by the unexpected. Mix up your executive DNA so you have bold thinkers, cautious planners, and everything in between. That diversity builds in resilience and keeps you steady through the storms.
- Trust conquers fear. Guide your people through AI-driven changes with honesty, training, and support. When employees

trust that you're looking out for them, fear turns into focus and engagement.

- Measure the human stuff. Don't just track sales and product metrics, track how your leadership system is functioning. If something's off in decision speed, alignment, or team health, address it like you would any mission-critical system.

- Look beyond the obvious. The next game-changing leader might not be the person with the fanciest title. Use data, dig deeper, and be willing to elevate someone who breaks the mold. Hallmark did it, and it transformed their future. Scouting for unorthodox talent can unlock huge opportunities.

"In the AI era, governing innovation
means governing human
capability."
— **Greg Brisco**

CHAPTER 9

❧

Community, Building a People-First Ecosystem

Innovation doesn't happen in isolation. You can have the best ideas and the sharpest technology, but without a community to amplify and support those ideas, they will wither on the vine. Community is the ecosystem that makes innovation sustainable, the safety net when experiments fail and the wind at your back when you push forward.

In this third pillar, we explore community on two fronts: inside your organization (your culture) and outside of it (your broader network of partners). Both are mission-critical in a Human-First, AI-Forward future.

Internal Community: Culture as Connection

Within an organization, *community* is another word for *culture*. It's the sense of belonging, trust, and shared purpose that inspires people to go beyond their job descriptions and give discretionary effort. Culture is what encourages a junior analyst to speak up with a bold idea because they know their perspective is valued. It's what allows a project team to admit a mistake without panic, knowing it will be treated as a lesson, not a liability.

Crucially, this kind of culture doesn't just materialize on its own, it's deliberately built. Leaders shape it through actions and rituals that knit people together. For example, you might host regular "innovation roundtables" where anyone from an intern to a VP can pitch ideas or flag opportunities. These aren't just feel-good sessions; they signal that curiosity matters and every voice counts. Another practice is two-way mentoring: pairing a senior leader with a tech-savvy rising star. The veteran gains fresh insights from the trenches while the younger talent learns leadership and strategy from decades of experience. Both sides grow, and the whole organization benefits as silos break down and hierarchy fades.

Here's the bottom line. When change hits, whether it's a new AI tool or a sudden market shift, your internal community will determine whether people panic or persevere. Teams grounded in real connection and psychological safety will pull together and lean into the challenge. But if people feel isolated or mistrustful, resistance sets in and progress grinds to a halt. Community is the glue that keeps morale high and agility intact when you need it most.

External Community: Networks and Partnerships

Your community doesn't end where your org chart does. In the AI era, no one succeeds alone. The organizations that thrive are those that build rich networks beyond their own walls. I'm talking about partnerships with universities, industry associations, startups, customers, even nonprofits. These external connections become force multipliers for learning and innovation.

For example, at Humanize Innovation we partnered with the Society for Collegiate Leadership & Achievement (SCLA) to extend our impact to the next generation. By integrating ISPI™ into SCLA's student leadership programs, we equipped a whole wave of young talent with a tool to discover their innovation strengths. Think about that ripple effect: a flood of new graduates entering the workforce already primed to be deeply human, endlessly adaptive, and radically inventive team players. That's exponential impact beyond any single company's reach, and it happened because we looked outside our own walls and chose to share our methodology generously.

The lesson here is straightforward. Investing in external community isn't charity; it's strategy. When you contribute to a broader ecosystem, you're planting seeds you'll later harvest. Those students reached through SCLA will fan out into a wide range of companies, maybe even yours, carrying forward the Human-First philosophy. The industry partnerships you cultivate mean you won't be scrambling alone when the next disruptive technology hits. You'll have allies, shared insights, and multiple perspectives tackling problems from different angles.

Membership Organizations: Signal in the Noise

Many of us belong to professional associations, chambers of commerce, or trade groups. These communities have always been about networking and knowledge sharing. But in an age of AI overload, their role has evolved. People today are drowning in information; what they crave is clarity. A membership organization can become a beacon by filtering the noise and delivering real insight.

For instance, rather than hosting yet another panel on AI buzzwords, a forward-thinking association might offer a workshop on practical AI ethics or introduce a framework like Innovation DNA Engineering™ as a member resource. In our experience, when associations began providing ISPI™ workshops and **Human First. AI Forward.** frameworks as member benefits, engagement shot up. Why? Because members gained something concrete: a career advantage, a new skill, a dose of clarity in a chaotic field. The association transformed from a social club into a catalyst for personal and professional growth.

And let's not forget the power of generosity. When leaders from different organizations come together in these communities and openly share what's working, whether it's their frameworks, successes, or even failures, it creates a culture of trust and reciprocity. Share your secret sauce, and others will share theirs. Over time, that trust compounds and expands everyone's capacity. You become known as a leader who isn't just in it for your own company, but for the greater good. That kind of reputation pays dividends in both opportunity and goodwill.

Community as Purpose

There's another dimension of community that transcends even networks and culture: purpose. This is where community and corporate responsibility intersect. In a Human-First approach to AI, we have a duty to ensure our innovations genuinely benefit people. It's not just about *what* we build, but *why* we build it and *who* we're helping along the way.

Companies leading the charge tie their innovation agenda to larger societal goals. Some establish initiatives to bring AI education and tools to underrepresented communities, shrinking the Human Delta™ and creating opportunity where there was little before. Others commit a share of their AI-driven gains to social causes or open-source projects. At Humanize Innovation, for example, we channel a portion of every engagement into our foundation, which funds community innovation projects ranging from scholarships for STEM students to grants for AI solutions tackling social challenges.

Why do this? Because it inspires your people and shows the world what you stand for. Employees feel proud knowing their work has a purpose beyond profit. Customers notice when a brand consistently acts on its values. And in a future where AI will reshape society, the companies that actively shape that future for the better are the ones people will want to work for, buy from, and invest in.

When you build community into your purpose, you elevate your organization from a market player to a movement leader. You're not only delivering products or services; you're rallying employees, partners, and even competitors around the idea that technology should serve humanity. That's the ultimate win-win, good for business and good for the world. And it cements your position as a truly **Human First. AI Forward.** leader.

Key Takeaways (Chapter 9):

Culture is your internal community. A strong culture of trust, safety, and collaboration makes your organization resilient in the face of change. It turns fear into focus.

No one innovates alone. External partnerships with academia, industry groups, nonprofits, and others can massively amplify your reach and insight. Your network is part of your innovation toolkit.

Clarity is key in chaos. In an information-saturated era, communities (like professional associations) that provide clear, actionable insight will win loyalty. Be the signal, not the noise.

Give to get. Generously sharing your knowledge and tools builds trust and attracts new opportunities. Generosity isn't just moral; it's a growth strategy.

Purpose powers community. When your innovation strategy visibly includes doing good, through education, equity, or ethics, you inspire employees and customers alike. You're not just running a business; you're leading a movement with a conscience.

"The leaders of tomorrow will not be
defined by what they know, but
by how human they remain in
an intelligent world."

— **Greg Brisco**

CHAPTER 10

❦

Education, Reimagining Education in the World of AI

Education isn't just another pillar in our Human First. AI Forward. framework, it's the foundation that supports all the rest. How we educate today will shape the next generation of leaders and prepare the workforce for an AI-driven future. The old educational model was built for a world of stability and predictability. It taught students to memorize facts, follow instructions, and color inside the lines. That made sense when information was scarce and change moved slowly.

But today, with AI reshaping every industry almost overnight, those old formulas are failing. The new era demands agility, creativity, emotional intelligence, and the ability to work across disciplines. In short, we need to reimagine education from the ground up to meet the demands of the AI age.

The Role of ISPI™ in Education

Think about how we traditionally introduce people to innovation tools. Usually, you don't encounter something like the ISPI™ (Innovation Strengths Preference Indicator) until well into your career, when a consultant hands your team an assessment to improve teamwork. Why wait that long?

As part of our Innovation DNA Engineering™ approach, we've started embedding ISPI™ into higher education. In partnership with the Society for Collegiate Leadership & Achievement (SCLA), college students map their innovation DNA before they even graduate. The idea is simple: give students an early look at their own strengths so they can make smarter choices. Are you a big-idea Pioneer or a detail-oriented Builder? More of a risk-taker or a risk-manager? Knowing that at 20 years old, instead of 40, allows a student to choose projects, internships, and career paths that truly fit who they are.

And it's not just about individual benefit. It creates a common language across campus. Suddenly professors, advisors, and students are all talking about innovation styles. A student team launching a startup on campus can have a conversation like, "We've got plenty of Pioneers on this project. Who can bring some Builder skills to balance us out?" This shared vocabulary strengthens collaboration, turning class projects and capstones into real-world training for innovation.

By the time these students enter the workforce, they're already fluent in how to build balanced teams and adapt on the fly. They've been trained not just in technical skills, but in how to design human-centered teams, a capability every bit as critical as coding or finance.

Why Education Must Change

AI has flipped the script on what it means to be "educated." In a world where any technical skill might be outdated in a year, the half-life of knowledge has never been shorter. What endures are the deeply human traits: adaptability, empathy, ethical judgment, creativity, and collaboration. These are the skills machines can't replicate and that only grow more valuable as automation expands. Modern curricula have to put these traits front and center.

Imagine a university where every student isn't just asked, "What's your major?" but also, "How do you innovate? What's your contribution style when you're tackling a tough challenge?" Picture seniors graduating not only with a degree in marketing or engineering, but with a personal playbook for innovation, a clear sense of how they can drive change and work with others in an AI-powered world. If higher education makes that shift, we won't just have graduates who can find a job. We'll have graduates who can reshape jobs and thrive in any environment where AI is pervasive.

Implications for Higher Education

What would it look like for universities to embrace this human-first, AI-forward approach? A few moves can make a big difference:

- Embed Innovation DNA Early: Introduce frameworks like ISPI™ from day one. Orientation programs, freshman seminars, and leadership workshops should weave innovation self-awareness into the fabric of campus life. Students should graduate fluent in their own strengths and in how to apply them.
- Design Balanced Teams in Class: Professors can use ISPI™ profiles to build project teams that mix thinkers and doers, Pioneers and Builders. When every class project becomes an exercise in balancing diverse skills, students learn team design as a core competency.
- Rethink Career Services: Instead of matching students to jobs based solely on majors, career counselors can guide students toward internships and roles aligned with their natural innovation style. For example, a student who knows she's a natural Builder might thrive in an operations-focused role at a startup rather than a pure R&D role, and a good advisor can spot that fit early.

- Partner with Employers: Companies can collaborate with universities to create pipelines of talent attuned to both technical and human skills. An employer might sponsor an innovation mentorship program or a case competition, gaining early access to students who understand their Innovation Readiness™ and can hit the ground running. It's a win-win: students find roles where they can excel, and employers gain young talent that's already innovation-ready.

A Vision for the Future

We've seen a glimpse of this future in our work with SCLA, a student leadership organization that spans hundreds of campuses and over 100,000 learners. It proves that education can be both massively scalable and deeply personal at the same time. Here's how it works: AI handles the heavy lifting in the background, curating personalized content, pairing students with mentors or teammates, and suggesting resources, while humans provide the soul. Coaches and mentors interpret the results, lead discussions, and nurture the kind of growth no algorithm can spark. Technology becomes a multiplier of human connection, not a replacement for it.

In this vision, the university experience is less about drilling students on facts (which AI can deliver in milliseconds) and more about building the muscles for agility, empathy, and creative problem-solving. Students practice adapting to new situations, working in diverse teams, and leading projects where outcomes are uncertain. They graduate not with a head full of memorized data, but with the capacity to learn, adapt, and invent on the fly.

In short, the leaders of tomorrow must be deeply human, endlessly adaptive, radically inventive, and fearlessly innovative. In other words, they have to be exponential by design.

Key Takeaways (Chapter 10):

- Education is the bedrock. It's the foundational pillar that supports innovation, leadership, and community by shaping the workforce of tomorrow.
- Start with ISPI™ early. Giving college students an ISPI™ assessment early on arms them with self-knowledge. They graduate innovation-ready, aware of their strengths and how to leverage them.
- Human traits are the X-factor. In an AI-saturated world, skills like adaptability, empathy, and creativity aren't just "nice-to-haves", they are the enduring edge that will set apart those who thrive. Technology will change, but being deeply human is a timeless competitive advantage.
- Human-first universities win. Schools that orient around human-first, AI-forward principles will produce graduates who are not just job-ready, but leadership-ready. These students won't just plug into the future job market; they'll shape it from day one.

"Extraordinary results often come
not from new resources, but from
redesigning how people
work together."

— **Greg Brisco**

CHAPTER 11

❧

Raytheon Innovation Challenge, Engineering Teams for 10× Output

Even the most advanced companies can stumble at innovation if their human systems aren't aligned. Raytheon, a global leader in aerospace and defense, found this out the hard way. The company launched an ambitious internal "Innovation Challenge" to crowd-source AI-driven project ideas from employees across the enterprise. The buzz was incredible, submissions poured in by the hundreds, and leadership set aside a hefty budget to fund the best ideas. By all accounts, it should have been a blockbuster success.

But when the results came in, they fell flat. Out of all those proposals, only two were approved for funding. And even those two struggled to gain traction. It was baffling. Did a company with tens of thousands of bright minds really have only two good ideas? Senior leaders began asking an uncomfortable question: was there a genuine creativity shortage, or did we design this process wrong? It was like standing in a field of dry grass and seeing only two sparks catch fire. Something was stopping the rest from igniting. With so much talent in-house, the problem clearly wasn't a lack of ideas. We suspected it was the system used to evaluate and develop those ideas that was broken.

Diagnosing the Problem

When our team came in to troubleshoot, we had a hypothesis: the human dynamics around the innovation challenge were out of balance. To test it, we had everyone involved take the ISPI™ assessment, including both the members of the selection committee and the team leads who submitted ideas. When the data came back, it immediately illuminated two major issues:

- Bias in the Selection Committee: The committee responsible for reviewing ideas was composed almost entirely of long-tenured, detail-driven experts. In ISPI™ terms, it was a room full of strong Builders, people who excel at process, precision, and risk management. Exactly who you'd want running a complex defense project, right? But the flip side was that, as a group, they were inclined to dismiss bold, unconventional ideas out of hand. Anything that looked too "out of the box" made them instinctively uneasy. In short, the gatekeepers were almost too cautious.

- Imbalance in the Idea Teams: On the other end, many of the teams that submitted proposals were skewed in the opposite direction. There were plenty of bold Pioneers brimming with vision and creativity, but they were light on practical execution skills. They came up with moonshot ideas, which is great, but they had few Builders or Executors to ground those ideas in reality. These teams essentially threw fireworks over the wall to the committee without a solid plan for what to do if one actually got lit.

In a nutshell, Raytheon's innovation pipeline was misaligned at both ends. The selection committee was filtering ideas through an ultra-conservative lens, while the idea generators were dreaming without enough grounding. No wonder most ideas never stood a chance: the system had too much caution at the top and too little at the bottom.

Re-engineering the System

Fixing this wasn't about telling individuals to "be more open-minded" or "be more practical", it was about structurally changing the system to inject balance. We recommended two major interventions:

- Remix the Committee: We helped Raytheon reconfigure the selection committee for the next innovation cycle by adding members with stronger Pioneer traits, people known for their creativity and appetite for new ideas. We included rising stars and fresh perspectives, not just the usual veterans. Just as importantly, we briefed the entire committee on the biases associated with their own profiles. We were explicit: "As a group, you tend to favor safe projects. Be aware of that so you don't accidentally choke off the next big thing." That awareness training primed them to give bold ideas a fair shot instead of reflexively shooting them down.

- Balance the Teams: We didn't want those visionary project teams to lose their spark, but we did want them to have a plan. For each of the most promising ideas, we coached the teams to fill their talent gaps. For example, one proposal came from a group of brilliant scientists with an AI cybersecurity concept. It was pure Pioneer energy with zero operational plan. We paired them with an operations manager from another division to help turn their vision into a workable project plan. Another team had a strong technical idea but no clear sense of how to pitch it to end users, so we brought in a marketing strategist to help them craft their story and define the use case. In essence, we helped these teams recruit a few Builders and Translators to join their cause, even if only as part-time advisors.

The guiding principle was straightforward: every team needed both visionaries and executors, risk-takers and risk-managers. We institutionalized that balance rather than leaving it to chance.

The Results

The difference in the very next innovation cycle was night and day. This time, the selection committee greenlit eight projects, a fourfold increase from the previous round. All of a sudden, ideas that would have been dismissed before were getting a thumbs-up. Why? Because now there were people in the room saying, "Wait, this one's bold, but it could be big. Let's explore it." At the same time, each idea had a balanced team behind it, able to speak credibly to execution concerns.

Those eight funded teams hit the ground running. Because we had helped them shore up their weaknesses, their proposals didn't die in committee or falter in early execution. Instead, they quickly moved into prototyping and implementation. In the span of a year, Raytheon went from nurturing two meager projects to a dozen active innovation initiatives. Several of these went on to full deployment in core business units, delivering real improvements to operations and even spawning new product lines.

One Raytheon executive captured the shift perfectly: "It feels like we removed an invisible chokehold on innovation." That chokehold had been the hidden biases and imbalances in their human system. By tweaking the people dynamics, specifically who evaluates ideas and how teams are composed, we unlocked a flow of innovation that had been there all along, just blocked by structural issues. Notably, we did this without adding a huge budget or hiring a slew of outsiders. We achieved a massive increase in output simply by rearranging the human pieces of the puzzle they already had.

Broader Implications

Raytheon's experience is a powerful reminder that when innovation is stuck, the answer isn't always "more ideas" or "more money." Often, it's about better alignment. An organization can have all the talent and budget in the world, but if the setup is off, if the human system is out of balance, very little will actually make it through the pipeline. In Raytheon's case, realizing that all the right ingredients were there, just in the wrong proportions, was a revelation. They didn't need to bring in a wave of outside creative geniuses to spark progress. They just needed to rearrange how their existing people were engaging in the process.

In fact, many companies likely have untapped tenfold output lying dormant, constrained by invisible organizational biases or outdated team-formation habits. The fix is often relatively cheap and cheerful: shuffle team compositions, broaden who gets to weigh in on ideas, and create awareness of bias. These human-centered tweaks cost virtually nothing compared to a major new R&D program, yet they can yield exponential returns. Before rushing to pour more funds into innovation, it's worth asking whether you're squeezing the most out of the talent you already have. In Raytheon's case, a few smart adjustments unleashed a torrent of creativity and execution that had been there all along, just waiting for permission to roar.

Key Takeaways (Chapter 11):

- Mix up the gatekeepers. A selection committee made up of like-minded, overly cautious experts will inevitably throttle bold ideas. Raytheon learned to diversify its evaluators so that breakthrough concepts actually get a fair shot.
- Great ideas need balanced teams. A visionary idea with no plan is as doomed as a flawless plan with no vision. The magic happens when you pair the dreamers with the doers, Pioneers with Builders, creative brains with operational muscle. Every innovation team should have both.
- Shine a light on bias. Even seasoned leaders can mistake "unfamiliar" for "unworkable." Training people to recognize that bias leads to more open-minded decision-making. Awareness is a cheap, effective antidote that can keep promising ideas from being prematurely dismissed.
- Huge ROI from human tweaks. Raytheon quadrupled its successful innovations without spending an extra dollar or making a single new hire, just by re-engineering the people dynamics. Sometimes the biggest unlock comes from rethinking how your existing talent is deployed, especially when budgets are tight.
- Bake it into the culture. After seeing the results, Raytheon made these changes permanent. Now every innovation cycle starts by ensuring committees have a mix of styles and teams self-balance their innovation DNA. Sustaining high-velocity innovation means making balanced human systems the norm, not a one-time fix.

"Capital accelerates ideas,
but people determine outcomes."
— **Greg Brisco**

CHAPTER 12

❦

VC Portfolio Operating System, Human-Centric Investment

Venture capital and private equity firms pride themselves on financial acumen, valuations, term sheets, market sizing, all that jazz. But let's be real: at the end of the day, the greatest lever for value in any investment isn't the money; it's the people. Investors often say, "We bet on the jockey, not just the horse," meaning they back teams as much as ideas. Research backs this up: in one major study, 95% of VC firms said the management team was an important factor in their decisions, and nearly half said it was the single most important factor.

Yet here's the irony: even though investors talk endlessly about the importance of teams, the way they actually evaluate and support those teams is often superficial. Due diligence on the people might consist of a few reference calls and some gut-feel judgments during the pitch. Résumés and past track records get a look, sure, but those only tell you what someone has done, not how they operate or how they'll behave under real pressure. The result is huge blind spots. A startup might look perfect on paper, with a great product, a big market, solid traction, but if the team dynamics are off, that company becomes a ticking time bomb in the portfolio.

We use the term Human Delta™ to describe this gap. It's the space between the visionary story founders pitch and the actual capacity of their team to deliver on it. In deal after deal, that Human Delta™ is where things ultimately succeed or fail.

Every startup or acquisition comes with a Human Delta™, whether investors recognize it or not. It's essentially a risk factor that grows or shrinks based on team alignment and capability. Many investors un-knowingly underestimate this. They assume that if the idea is good and the market is big, throwing capital or a bit of "management guidance" at the company will iron out any team wrinkles. In reality, as a company grows, any misalignments in the team tend to expand.

Consider a founding team made up entirely of Pioneers. These vision-ary entrepreneurs can sell the dream and crank out 100-hour weeks to get something off the ground. In the early stages, that can work phenomenally, they'll create a bang and gain traction. But fast-forward a year or two: now they have to build processes, scale operations, and manage a growing headcount. Without any Builders (the execution and process gurus) on the leadership team, these founders will hit a wall. Execution stalls, details slip through the cracks, and frustration builds because their natural strengths don't include turning chaos into order. The Human Delta™ in this case is an execution gap, a gulf between the vision and the ability to operationalize it.

On the flip side, imagine a company run by cautious, methodical types, let's call them Adapters in our framework. They nail execu-tion and keep pristine books. Early growth might be steady, even boring (in a good way). But then the market shifts or a disruptive opportunity emerges. This team may lack the bold, risk-taking Pioneer mindset needed to seize it. Their company could pla-teau or get outflanked by a more aggressive competitor. Here, the Human Delta™ is a boldness gap. The team can deliver on what they know, but they struggle to reinvent themselves when the game changes.

The frustrating part is that these patterns, too many Pioneers and too few Builders, or the reverse, are often visible from the start if you know how to look. But traditional due diligence doesn't systematically measure these dynamics. As a result, investors often recognize the problem only in hindsight, after a portfolio company falters.

Using ISPI™ in Due Diligence

How can we bring more structure and foresight into assessing teams before we write the check? One tool in our playbook is to have founders and key executives complete the ISPI™ as part of due diligence (with trust and permission, of course). Think of it as an Ultrasound of the team's innovation DNA. The patterns that emerge can be incredibly telling for an investor:

- If an executive team's ISPI™ results come back as all extreme Pioneers, that signals execution risk. Before cutting the check, you know these leaders will need serious operational support sooner rather than later. That might mean planning to hire a seasoned COO or strengthening middle management with experienced operators. It doesn't mean you don't invest, it means you invest with eyes wide open about what's likely to break first.
- If instead you find a team of uniformly cautious Adapters, great at optimization but low on big-vision energy, you've got the opposite issue. This company might run like a Swiss watch in the short term, but you can predict it will struggle to pivot or pursue disruptive innovation. As an investor, you might decide, "We'll fund them, but we need to pair them with a visionary advisor or install a Pioneer-minded board member to encourage bolder thinking."

The ISPI™ gives a common language to discuss these dynamics without it feeling like a personal critique of the founders. It turns a gut feeling into data. You're effectively saying, "This is the innovation profile of the team. Does that profile match the strategy this company will need as it grows?"

It also makes conversations with founders more productive. Instead of vaguely saying, "We have some concerns about the team," you can be specific: "It looks like you don't have a natural Builder in your core group. Let's talk about how you're handling operations as you scale." Many founders actually appreciate this insight, imagine an investor who is taking a genuine, data-informed interest in their team's success, not just scrutinizing the revenue curve. It signals that you're not looking to blame the team, but to support it.

Post-Investment Alignment

The best investors don't just diagnose team issues, they help fix them. After the deal is done, there's often a huge opportunity to add value by aligning and coaching the team before things go off the rails. Unfortunately, many investors default to a heavy-handed approach when a portfolio company hits turbulence: they pressure the founders, or even start considering replacing them. But swapping out leadership can be like doing a heart transplant on a sprinting athlete, it's incredibly risky and costly for everyone involved.

A far better move is to engage early with human-centered interventions. I'll share a real example. A private equity firm we advised acquired a company where, about a year in, the visionary founders were clashing with a newly hired operations team. It was oil and water. The founders felt the ops leaders were bureaucratizing their "baby." The operations team felt the founders were unpredictable and undermining the structure they were trying to build. The board was getting nervous; conversations of "maybe we need a new CEO" had started.

We stepped in with an Innovation Alignment workshop. First, we laid everyone's ISPI™ profiles on the table, which immediately deperson- alized the conflict. Sure enough, the founders were hardcore, indepen- dent Pioneers, and the operations team members were collaborative Builders. Neither group was "wrong", they simply hadn't realized they were speaking different languages.

In the workshop, we helped both sides see these differences clearly and appreciate the value in each other's approach. We facilitated frank discussions to reset expectations. Which decisions would the founders delegate to the operations team? How could the operations leaders adapt to the founders' need for autonomy in creative work? We es- tablished new norms and roles that played to each group's strengths instead of pitting them against each other. Over the next few months, tensions eased dramatically. Collaboration improved, shaky project milestones got back on track, and the company started growing again without any executive casualties.

That intervention likely saved millions, not only by avoiding the cost and disruption of a leadership shake-up, but by preserving trust and relationships with the founding team. Equally important, it changed how the private equity firm was perceived. Instead of being seen as number-crunchers ready to axe the team, they earned a reputation as partners who help you get it right. They became known as a firm that doesn't just invest capital, but also invests in aligning the human en- gines of its portfolio companies.

Branding Advantage for Investors

In the venture and PE world, reputation is a currency of its own. The best entrepreneurs often have multiple funding options, and they talk to each other. If you build a brand as a human-first investor, it becomes a magnet for top-tier opportunities. Imagine the word on the street: "This VC won't just flood you with spreadsheets and growth demands. They actually help you build your team and culture. They understand the people side of the business." Founders will seek out that investor because they know the partnership will be about more than cash, it will involve mentorship, support, and a better shot at long-term success.

We're already seeing this shift. Some forward-thinking investment firms are hiring people with titles like Talent Partner or Platform Manager who focus on helping portfolio companies with recruiting, culture, and organizational development. That's a step in the right direction, but the real differentiator is the underlying ethos. When a firm truly believes in Human First. AI Forward. meaning they put people at the center while leveraging the best technology and data, it shows in their actions. They run Innovation DNA sessions at off-sites with new startup teams. They connect founders with executive coaches before problems arise. They encourage cross-pollination of ideas across the portfolio so CEOs can learn from each other's mistakes and successes.

This approach does more than support existing investments; it attracts fearlessly innovative founders who want that kind of partnership. In a crowded capital market, being known for deep people insight and a genuine commitment to founder success beyond the term sheet is a powerful edge. It's like being the VC firm with a soul. And in an era where money has become a commodity, having a soul is a serious competitive advantage.

Key Takeaways (Chapter 12):

- Diligence the team, not just the numbers. Evaluating the people leading a company is just as critical as vetting the financials or the technology. Tools like ISPI™ give investors a structured view into a team's working style and capacity, revealing alignment or misalignment before they invest.

- Intervene early, don't just replace. When a portfolio company stumbles, avoid the knee-jerk reaction to swap out founders or execs. Often a bit of targeted coaching, role realignment, or conflict mediation can fix the issue faster and with far less collateral damage than a leadership coup. Early intervention can preserve both the talent and the original vision that made you invest in the first place.

- Build a human-first reputation. In venture capital, being known as the firm that genuinely helps founders build great teams and healthy cultures is pure gold. The best entrepreneurs will choose investors who are true partners in growth, not just financial backers. Brand yourself as the ally who "gets the people side," and you'll draw more fearlessly innovative founders into your orbit.

- Human systems drive scale (or stall it). As companies scale up, tiny team misalignments can snowball into big problems. A team that's "good enough" at $1M in revenue might crack at $50M. By proactively engineering alignment, making sure the right people are in the right roles with the right chemistry, investors can de-risk growth. Fix those hairline fractures in the human system before they become gaping cracks.

- Blend financial savvy with people savvy. The investors of the future are part financial engineer and part human-systems engineer. By combining traditional deal-making skills with innovation coaching and team design, firms can create far more value than either approach alone. This cross-pollination of finance and human insight leads to outcomes that are truly exponential by design: better returns, stronger companies, and lasting partnerships.

"Innovation stops being a department
the moment everyone sees
themselves as part of it."
— **Greg Brisco**

CHAPTER 13

❦

Shared Responsibility, Aligning Stakeholders for AI Success

Aligning people for innovation is not just an HR initiative or a leadership-training checkbox. It's a boardroom issue and a daily management mandate. In an era of constant disruption, strict regulations, and fierce competition, innovation isn't optional, it's survival. Yet roughly 70% of digital transformations still fail, and the majority falter for one simple reason: people. Boards often sign off on multi-million-dollar AI projects without scrutinizing the human dynamics underneath, essentially approving high-risk experiments poised to collapse.

Even the smartest strategy with the biggest budget will crash without the right people aligned behind it. Innovation succeeds when everyone, from directors in the boardroom to employees on the front lines, owns a piece of the effort. This is the essence of Innovation Readiness™. It shifts the focus from asking whether the technology is ready to asking whether people are ready to leverage it.

Let's break down what this shared responsibility looks like at each level of the organization:

Boards: Setting the Tone

Traditionally, corporate boards have obsessed over finances, compliance, and picking the right CEO. Those responsibilities remain, but today's boards must also become guardians of innovation capacity and culture. A company might look stable on a balance sheet yet be one big shake-up away from crumbling because its leadership culture is too cautious, too homogeneous, or not wired for bold innovation.

Forward-thinking boards now ask themselves tough, pointed questions:

- Do we have bold Pioneers in key roles if our strategy demands breakthrough ideas?
- Do we have resilient, change-agile Builders in place if our market becomes volatile?
- Do we have the right mix of risk-takers and risk-managers to match our vision?

These aren't "soft" questions, they're fundamental to governance in the AI era. ISPI™ data can give boards a dashboard for these human factors. Instead of relying on gut feel or rosy management reports, directors can see hard metrics on the organization's innovation DNA. They can pinpoint where talent gaps exist, where the culture may be too risk-averse, and where groupthink might lurk. Armed with that insight, a board can push leadership to address real people-related blockers to innovation before green-lighting the next big initiative.

The CEO: Architect of Alignment

If the board sets the mandate, the CEO is the architect who must bring it to life. CEOs carry the weight of aligning vision with execution. Too often, a chief executive unveils an ambitious, cutting-edge strategy and simply assumes the teams will "get there." Meanwhile, the gap between aspiration and reality, the Human Delta™, yawns wide beneath them.

Great CEOs close that gap proactively. They use tools like the ISPI™ to reveal whether their leadership team skews too heavily toward operators at the expense of visionaries, or vice versa. Armed with that data, they can course-correct, reshuffling roles, bringing in new talent, or doubling down on development, long before execution goes off the rails.

Culture, after all, is not an accident; it's engineered. The most effective CEOs treat culture-building as Innovation DNA Engineering™. They deliberately design teams and norms to be adaptive rather than reactive. They embed the traits of a high-performing, innovative organization, deeply human, endlessly adaptive, fearlessly inventive, so the company becomes exponential by design rather than by luck. In turbulent times, this kind of intentional culture is what separates companies that thrive from those that falter.

CHROs: Engineers of Human Capacity

Chief Human Resources Officers have a critical role in this shared model. Gone are the days when HR was confined to hiring, firing, and annual training sessions. In an AI-driven world, CHROs must become true engineers of human capacity and culture.

By applying the ISPI™ across recruitment, leadership development, and succession planning, HR can give the CEO and board evidence-based answers to the ultimate question: "Do we have the people and culture to win?" With ISPI™ profiles in hand, a CHRO can pinpoint whether the next generation of leaders has what it takes to execute the company's strategy, well before they ever step into the C-suite. They can identify which teams need more diversity in thinking styles, where innovation hot-spots exist, and who might struggle in a high-change environment.

When HR leaders bring this level of strategic insight, they stop being a support function and start becoming a driving force. They move from the periphery of strategy to its center, ensuring the company's growth plans are grounded in human reality.

Technology Leaders: Bridging Systems and People

For CIOs and CTOs, the leaders steering major technology investments, the challenge of AI isn't the tech itself. It's getting people to actually use it. History is littered with powerful systems that failed simply because employees resisted change or didn't see "what's in it for me."

Technology leaders can bridge this gap by pairing ISPI™ insights with their rollout plans. Say you're deploying a major AI-powered platform. Using ISPI™ data, a CIO can identify who in the organization are natural early adopters and who might be skeptics. They can recruit the innovators and early adopters as change champions to lead by example, while arranging extra support and training for the more cautious or change-resistant. In short, they orchestrate the rollout as a human adoption campaign, not just a technical installation.

When tech leaders approach implementation with this people-first mindset, adoption stops being a coin toss. New systems stick because they're embraced by the culture rather than force-fit. The result: technology investments actually deliver the competitive edge they promised, because the people are on board.

Strategy Executives: Closing the Execution Gap

For the EVPs and SVPs tasked with strategy, a brilliant plan can die on the vine if execution falters. And execution usually falters not because the plan was flawed, but because people weren't aligned in carrying it out.

Armed with ISPI™ data, strategy leaders can assemble cross-functional teams with the right balance of thinkers. Need a moonshot initiative to diversify the business? Make sure you've got enough bold Pioneers in the mix. Launching an efficiency drive? You'll want that team stacked with detail-oriented Builders. Using this common language of innovation styles, strategy executives can even predict where friction might arise. For example, if they notice a project team has a pair of big-picture visionaries working alongside a group of cautious, process-driven guardians, they can anticipate tension. Instead of waiting for conflict, they might adjust the lineup or set clear ground rules upfront.

In this way, strategy executives help close the gap between big ideas and ground-level execution. Plans don't just live on PowerPoint slides, they get implemented and deliver real results.

Mid-Level Leaders: Building the Next Generation

Directors and senior managers in the middle of the organization form the leadership pipeline that will steer the company tomorrow. For them, embracing these alignment tools isn't just about building better teams; it's about personal growth. When mid-level leaders use the ISPI™ to understand their own innovation strengths and blind spots, it's like receiving a personalized roadmap for career development.

Armed with that self-awareness, they can intentionally stretch beyond their comfort zones. A manager who learns she's a strong Builder but a weaker Pioneer might seek out more visionary projects or pair with a Pioneer mentor to develop that muscle. These leaders also use such insights to construct complementary teams beneath them, filling roles with people who balance their own tendencies.

The result is twofold: Teams perform better and these rising leaders accelerate their readiness for higher roles. The leaders who practice alignment and adaptability today naturally become the executives of tomorrow.

Every Employee: Owning Their Role in Innovation

Innovation isn't the sole province of R&D or a special "innovation lab." In a Human-First, AI-Forward organization, every employee has a stake in pushing the company forward. And every employee, whether an intern or a senior specialist, has their own innovation strengths.

When people at all levels become aware of their ISPI™ profiles, it's empowering. Suddenly a junior analyst who once felt out of place realizes she's actually a natural Pioneer who can offer bold ideas. An engineer discovers he's a crucial Stabilizer when chaos hits. Armed with this self-knowledge, employees can align their daily work to their strengths instead of grinding against the grain. Frustration drops, engagement rises, and great ideas start coming from everywhere.

Just as important, when employees see leadership investing in these human-centered tools and openly talking about alignment, they feel valued rather than managed. It creates trust and a sense of shared mission. The message becomes clear: we're all in this innovation journey together. That feeling of ownership at the ground level becomes jet fuel for innovation, and a powerful antidote to turnover.

Key Takeaways (Chapter 13):

- Alignment is everyone's job. Driving innovation is a shared responsibility stretching from the boardroom to the front lines. Every level has a role to play in closing the Human Delta™.
- Boards must govern for innovation DNA. Modern boards treat people-risk as real and measurable. They insist on seeing

the company's innovation DNA and culture health right alongside the financials.

- CEOs have to design culture on purpose. Culture isn't going to fix itself. Top executives must intentionally build cultures that are ready for change. They practice their own form of Innovation DNA Engineering™ to make the organization adaptive by design.
- HR and tech leaders are strategic linchpins. CHROs and CIOs (and CTOs) aren't back-office players anymore. They drive strategy by ensuring the right people are in the right seats and that new technologies are embraced, not resisted.
- Mid-level leaders accelerate the pipeline. When rising leaders use tools like ISPI™ to grow themselves and craft better teams, they turbocharge both current execution and their future careers.
- Every employee contributes to innovation. Every single person has innovation strengths that matter. When those strengths are recognized and channeled, innovation stops being a buzzword and becomes everyone's everyday business.

"When we lead with humanity,
technology finally learns
how to follow."

— **Greg Brisco**

CHAPTER 14

❧

Responsible and Ethical AI

Ensuring AI Serves Humanity's Best Interests

As artificial intelligence becomes increasingly embedded in business and society, the need for responsible and ethical AI has never been more critical. AI is a powerful tool with tremendous potential, from automating tasks and informing decisions to augmenting human creativity, but without proper oversight it can also perpetuate biases, violate privacy, or make harmful mistakes. A bruised reputation, stakeholder divestment, and even legal liability are real risks when AI is used without appropriate governance. Ultimately, human beings remain accountable for AI outcomes, so we must guide AI development and deployment with a strong ethical compass. This chapter explores what "responsible AI" entails, why it matters, and how Humanize Innovation is intertwined with the entire concept, ensuring that technology always remains aligned with human values and ingenuity.

What Does "Responsible AI" Mean?

In simple terms, responsible AI refers to the principles and practices that ensure AI systems are developed and used in a manner that is ethical, transparent, and accountable. It is about deliberately guiding AI so that it benefits people and minimizes harm. Experts describe responsible AI as paying careful attention to fairness, addressing biases, and continually checking an AI system's impact on all stakeholders. While the terms "ethical AI" and "responsible AI" are often used interchangeably, they carry slight differences in emphasis. Ethical AI is a broad, philosophical approach focused on high-level principles (like justice, autonomy, and societal impact), whereas responsible AI typically refers to the concrete, day-to-day implementation of those principles in organizations. In practice, both concepts overlap, a responsible AI strategy is how an organization operationalizes ethical considerations, ensuring its AI initiatives uphold core values and do not undermine its mission.

Put another way, responsible AI is about building trust. For people to trust AI systems, they must be confident that these systems are doing the right thing, making fair decisions, respecting privacy, and being transparent about how they work. As IBM describes it, responsible AI involves aligning AI with stakeholder values, legal standards, and ethical principles, embedding those principles into AI design and use to maximize positive outcomes while mitigating risks. When AI is trustworthy, it augments human capabilities instead of eroding them. When it's not, the consequences can be severe, from biased hiring algorithms to unsafe autonomous systems. A now-famous example is Amazon's experimental recruiting AI that had to be scrapped after it was found to systematically discriminate against women applicants due to learning from biased historical data. Such cautionary tales underscore why proactive ethical guidelines are paramount.

Principles of Ethical AI

Over the past few years, industry and academic leaders have converged on a set of core principles that define responsible, ethical AI. According to one Harvard framework, organizations that use AI ethically tend to adhere to five key pillars: fairness, transparency, accountability, privacy, and security. In essence, this means:

Fairness, Strive to eliminate bias and discrimination in AI outcomes. A fair AI system's decisions should be equitable across all groups, avoiding unjust impacts on protected classes (e.g. not systematically disadvantaging people of a certain gender or ethnicity). Achieving fairness may require carefully curating training data, applying bias mitigation techniques, and continually testing outputs for disparate impacts.

Transparency, Be clear and open about how AI systems operate. Transparency involves knowing what data goes into an algorithm and how the algorithm makes decisions. This could mean making aspects of the AI's logic explainable to users or providing documentation about its design. Transparent AI builds trust because people can scrutinize and understand its reasoning.

Accountability, Establish human oversight and responsibility for AI behavior. Even when decisions are automated, an accountable AI framework ensures that people (with names and roles) are answerable for those outcomes and can intervene when needed. Mechanisms like audit trails, AI ethics review boards, and clear escalation paths help enforce accountability.

Privacy, Safeguard personal data used by AI. Ethical AI respects user privacy by minimizing data collection to what's necessary, securing that data, and complying with regulations. Techniques like anonymization and federated learning can help protect sensitive information while still enabling AI insights.

Security, Protect AI systems from misuse, tampering, and cyber threats. AI models and data must be secured against hacking or manipulation, which could otherwise lead to dangerous outputs. Security also means reliability and safety, ensuring the AI behaves as intended even under unusual or stressful co

These principles are widely echoed across the AI community. For example, Microsoft's responsible AI guidelines include very similar themes, fairness, reliability & safety, privacy & security, inclusiveness, transparency, and accountability, as key considerations for any AI, especially generative AI. By adhering to such pillars, organizations can develop AI systems that are not only effective, but also worthy of trust. The goal is to maximize AI's benefits while minimizing its risks, enabling innovation to flourish without compromising our ethical standards.

Why Ethical AI Matters

Focusing on ethics in AI is not just altruism; it's essential risk management and a driver of sustainable success. AI systems, by their nature, learn from historical data and make autonomous decisions, which means they can inadvertently amplify human biases or make opaque choices that impact lives. Without ethical guardrails, AI could undermine fundamental values and even a company's mission. "Taking a responsible approach to AI allows businesses to deliver results without undermining their mission and values, or exposing them to litigation," notes Michael Impink, an AI ethics instructor. In highly regulated arenas like finance or healthcare, failures of responsibility can lead to breaches of law and public trust. Even in less regulated spaces, consumer backlash can be swift if an AI product is seen as creepy, unfair, or unsafe.

Moreover, ethical lapses hurt the effectiveness of AI. Biased or erroneous AI decisions erode user confidence and can prompt good employees or customers to disengage. Conversely, organizations that champion responsible AI can differentiate themselves and gain a competitive edge. Leaders well-versed in AI ethics will spot potential issues faster and steer their teams to fix them, avoiding costly mistakes. They will also be better positioned to navigate emerging regulations (such as the EU's AI Act) and to shape positive public narratives around AI. In short, ethical AI is about ensuring technology serves humanity's best interests. It's a commitment that whatever incredible capabilities we build with AI, we will continually align them with human rights, social values, and the greater good.

Humanize Innovation embraces this commitment wholeheartedly. As the AI revolution accelerates, we insist that progress must not come at the expense of our humanity. Each innovation must be evaluated not only by what it achieves, but how it achieves it and whom it impacts. In the next section, we'll discuss how Humanize Innovation brings a uniquely human-centric approach to AI development, illustrating how responsible AI principles are woven throughout our methodology.

Humanize Innovation: Putting Humans at the Center of AI

Humanize Innovation's mission is to ensure that innovation begins and ends with people, even when that innovation is powered by AI. This perspective is deeply intertwined with responsible AI. We recognize that no matter how advanced algorithms become, humans must remain in charge of defining the purpose, parameters, and ethical boundaries of technology. In practice, that means Humanize Innovation emphasizes collaboration between human ingenuity and AI at every step. We advocate for "human-in-the-loop" approaches where AI augments human decision-making instead of replacing it outright. By keeping skilled people involved to supervise AI outputs, provide feedback, and inject moral judgment, we reduce the risk of AI running amok or making unsound choices.

Our approach also leverages tools and frameworks that reveal the human side of innovation, ensuring technology implementations align with human strengths and values. For instance, the Innovation Strengths Preference Indicator (ISPI™), a tool developed with our partner Idea Connection Systems, helps organizations map out how different people think, create, and collaborate. By using ISPI™ alongside AI deployments, teams can configure AI roles that complement human team members' creative strengths, rather than conflict with them. This is an example of how Humanize Innovation blends behavioral science with technology: we make the "invisible" human dynamics visible and measurable so that AI solutions can be tailored to fit the people they serve. A responsible AI project in our view is not just about the code or data; it's about the culture and context in which the AI operates. We help clients build innovation cultures where ethical considerations, like inclusivity and accountability, are baked into project goals from the start, not bolted on as an afterthought.

Crucially, Humanize Innovation champions diversity, equity, and inclusion (DEI) as key tenets of ethical AI. Just as diverse teams produce better outcomes in general, diverse voices are vital in AI development to catch biases and blind spots. We strive to bring together interdisciplinary teams (engineers, ethicists, domain experts, end-users) to co-create AI solutions. A mosaic of perspectives ensures that AI tools are vetted from multiple angles and work well for different populations. This philosophy aligns with emerging best practices; for example, IBM advises involving interdisciplinary and diverse teams in AI development to help identify biases that homogeneous teams might miss. By mirroring this in our projects, including both technical and non-technical stakeholders in reviewing an AI model's results, we foster fairness and broader acceptability of AI innovations.

Finally, Humanize Innovation is committed to education and transparency around AI. We believe that demystifying AI, explaining in plain language what it can and cannot do, empowers people at all levels of an organization to engage with it responsibly. In our workshops and "Humanity Labs," we frequently include sessions on AI ethics and literacy, ensuring leaders and teams understand concepts like algorithmic bias, model explainability, and data privacy. Our goal is to cultivate ethical leadership in the AI era. That means equipping decision-makers with both the technical insight and moral courage to question how AI is used. We often invoke the guiding principle: "Just because we can do something with AI does not mean we should." Humanize Innovation encourages organizations to define clear usage policies (e.g. no AI use that violates our core values), establish ethics committees for AI projects, and create feedback channels for employees or customers to raise concerns about AI systems.

By intertwining these human-centered strategies with technical innovation, Humanize Innovation ensures that responsible and ethical AI isn't just a checkbox, but a living practice. As we partner with companies, universities, and communities, we continuously reinforce the idea that innovation has a pulse, a human heartbeat that technology should never drown out. In summary, responsible AI is not a hurdle to innovation; it is the foundation of sustainable innovation. Guided by ethical principles and human insight, we can harness AI in ways that amplify our ingenuity, advance our well-being, and uphold our values. Humanize Innovation stands at this very intersection, proving that when we humanize AI, we unlock its most transformative and positive possibilities.

(With the core concepts of Responsible AI explored, we now turn to acknowledging the partners who have joined Humanize Innovation in championing this vision. The following section highlights our strategic corporate partnerships and how each organization's mission aligns with our human-centered approach to innovation.)

Corporate Partnerships and Aligned Missions

Humanize Innovation is not alone in the quest to humanize technology and foster ethical innovation. We are proud to collaborate with a select group of strategic partners, organizations whose expertise and values align closely with our mission. Each partner brings a unique strength to the table, from cultivating innovative mindsets to developing principled leaders. Below, we introduce our corporate partners and discuss how each contributes to the shared vision of innovation that is inclusive, ethical, and human-centered.

Idea Connection Systems (ICS)

Idea Connection Systems, Inc. (ICS) is a pioneering innovation consultancy that has been "bringing innovation to the world" for over 30 years. Founded by Bob Rosenfeld (who previously led innovation at Eastman Kodak), ICS specializes in understanding the human dynamics of innovation. They teach methods and create tools to measure and facilitate innovation within individuals, organizations, and communities. A hallmark of ICS's approach is the recognition that innovation does not happen by accident or decree, it arises from aligning the right people with the right projects under supportive leadership. Notably, ICS developed the Innovation Strengths Preference Indicator (ISPI™), a psychometric tool that reveals how different people prefer to innovate and collaborate, thus helping teams to capitalize on their collective strengths.

Alignment with Humanize Innovation: ICS's purpose is "to encourage a better future by stimulating the growth of an ever-advancing civilization" and its mission is "to partner with clients in their creation of the new, leveraging awareness of different mental models to make innovation happen". This philosophy is a natural fit with Humanize Innovation. Both ICS and Humanize believe that people are at the heart of every successful innovation. Our collaboration leverages ICS's human-centric tools like ISPI™ to ensure that when we introduce AI or other technologies, we do so in harmony with human creativity and not at its expense. By quantifying traits like how individuals approach problem-solving or change, ICS helps inform our tailored strategies for AI integration, e.g., identifying innovation champions and ensuring diverse thinking styles are represented on a project. Moreover, ICS's values (which include trust, integrity, mutual responsibility, and continuous learning) echo the ethical underpinnings of responsible AI. Together, Humanize Innovation and ICS co-create environments where technological innovation is guided by empathy, inclusion, and a deep understanding of human potential.

Society for Collegiate Leadership & Success (SCLA)

The Society for Collegiate Leadership & Achievement (SCLA), referred to in our partnership as the Society for Collegiate Leadership & Success, is a nationally accredited honor society that helps college students "build a career that matters." SCLA's mission is to honor academic achievement and prepare students for success by bridging the gap between college and the professional world. Through its presence on over 850 campuses and a 100,000+ member network, SCLA provides career-readiness programs, leadership development, and access to real-world opportunities at scale. What makes SCLA particularly innovative is its use of technology and data to empower students: members benefit from an AI-powered career hub with résumé builders and smart chat advisors, personalized skill development plans, and even an online mentorship matching system, and microinternship partnerships. By combining honor society recognition with practical career tools, SCLA ensures that students not only celebrate their collegiate success but also translate it into tangible post-graduate outcomes. For example, SCLA aligns its programming with recognized career readiness competencies and provides accredited certification programs, positioning its members a step ahead in the job market.

Alignment with Humanize Innovation: SCLA's work is fundamentally about investing in human potential, which sits at the core of Humanize Innovation's ethos. By preparing the next generation of leaders and innovators, SCLA addresses a key piece of the ethical AI puzzle: ensuring that future decision-makers have the leadership skills and moral grounding to use technology responsibly. Humanize Innovation partners with SCLA to infuse cutting-edge insights on human-AI collaboration into collegiate programs, while also learning from SCLA's expertise in scalable learning communities. For instance, SCLA's emphasis on experiential learning and measurable outcomes for students resonates with our approach to continuous innovation training within organizations. We both champion lifelong learning, adaptability, and purpose-driven leadership. SCLA's motto, "Empowering you to build a career that matters", aligns with our message that AI and innovation should be channeled toward meaningful, positive impact. Together, we envision a future where young professionals enter the workforce not only tech-savvy but also "human-savvy," equipped to maintain empathy, ethics, and creativity alongside the most advanced AI tools.

Stedman Graham & Associates, Identity Leadership

Stedman Graham & Associates (SGA) is a leadership development and consulting firm headed by Stedman Graham, a bestselling author and educator known for his groundbreaking work on "Identity Leadership." The core of SGA's philosophy is captured in Graham's mantra: "To lead others, you must first be able to lead yourself." Identity Leadership is a curriculum of self-discovery and personal strategy, it teaches that true leadership starts with knowing who you are, defining your purpose, and organizing your life around your core identity. Graham's Identity Leadership 9-Step Success Process is a structured journey that includes steps like "Check Your ID" (understand your strengths, influences, and values), "Create Your Vision," and "Build Your Dream Team," among others. By following this process, individuals learn to chart a course for their lives and careers that is authentic and principled. SGA delivers this content through workshops, keynotes, and online courses, empowering leaders from students to executives to take charge of their own development and destiny. The underlying message is one of personal accountability and continual growth: You are not your circumstances, but your possibilities, as Graham often says. In the context of organizational consulting, SGA helps companies cultivate leaders who have a strong sense of self, which in turn makes them more effective at leading teams and adapting to change.

Alignment with Humanize Innovation: The partnership between Humanize Innovation and SGA is built on the shared belief that human leadership is the linchpin of responsible innovation. No AI or advanced technology can compensate for a lack of vision or integrity at the top. Stedman Graham's Identity Leadership framework provides a powerful foundation for the kind of ethical, self-aware leadership that we believe should guide AI initiatives. By encouraging leaders to clarify their identity and values, SGA ensures that when those leaders implement AI or drive innovation, they do so anchored by a clear moral compass. Humanize Innovation integrates Identity Leadership principles into our programs to help leaders answer key questions before they deploy AI: What do I stand for? What impact do I want this technology to have? This introspective approach helps prevent the common pitfall of chasing innovation for its own sake. Instead, leaders first define success in human terms (such as improving customer well-being or employee growth), then use AI as a means to that end. Additionally, SGA's focus on education and empowerment complements our emphasis on training and culture. Whether we are working with a university cohort or a corporate C-suite, introducing Identity Leadership concepts amplifies the human-centric perspective, reinforcing that leading in the age of AI starts with leading oneself responsibly. Our collaboration with SGA and Stedman Graham helps clients build not just technical proficiency, but the strength of character and clarity of purpose required to navigate the complexities of modern innovation.

Collectively, these corporate partners form an ecosystem of expertise that bolsters every aspect of Humanize Innovation's mission. Idea Connection Systems contributes methods to measure and drive human-centered innovation; SCLA builds the pipeline of future ethical innovators; Stedman Graham & Associates infuses deep leadership wisdom and personal accountability; and Rapport Leadership ignites the human potential in every leader and team. Each partner shares our conviction that innovation must elevate people and that technological advances require equally advanced growth in human capability. By collaborating so closely, we are able to cross-pollinate ideas and co-create programs that none of us could deliver alone, programs that integrate technology with humanity, data with empathy, and AI with moral insight. In the next section, I (Greg) will offer some personal acknowledgments to these partners and others, reflecting on how they have shaped this journey. But first, a heartfelt thank you to all our partners for believing in the vision of humanized innovation and for working hand-in-hand with us to make it a reality.

Final Reflections and Acknowledgments

Writing Humanize Innovation – Human First. AI Forward has been a deeply personal journey. It is as much about my own growth and learning as it is about the revolutionary technologies reshaping our world.

I could not have completed this manuscript without the support, wisdom, and contributions of many individuals and organizations who share my passion for human-centered innovation. I'd like to take a moment to acknowledge them in my own voice, expressing my genuine gratitude and highlighting how each has aligned with and amplified the mission of Humanize Innovation.

First and foremost, I want to thank my corporate partners, the kind of partners every visionary initiative dreams of having.

Bob Rosenfeld and the team at Idea Connection Systems, you have shown me that innovation indeed has a heartbeat. Your decades of work proving that innovation thrives when we understand people's creative DNA has profoundly influenced Humanize Innovation's approach. The ISPI™ tool you developed gave us a lens to see the invisible forces of innovation within teams, and I am grateful for how we've applied it together. More than any theory, seeing how ICS lives its values of trust, integrity, and continuous learning inspired me to embed those same values into every AI project I undertake. Thank you for believing that a tech future still needs the human touch, and for teaching us how to measure and unleash that human potential in tandem with AI. Our mission to humanize AI is so much stronger with ICS's insight that innovation begins and ends with people.

To the Society for Collegiate Leadership & Success (SCLA), especially the leadership who welcomed me into your national advisory conversations, thank you for your relentless commitment to empowering young leaders. Working with SCLA has been like looking into the future and feeling very hopeful. You've built a platform that not only honors academic success but also arms students with practical skills and moral grounding. In a world worried about AI displacing jobs, you focus on humans replacing doubt with confidence. I've been moved by how SCLA integrates AI tools (like your AI career hub) in a student-friendly, ethical way, always with an educational purpose and never forgetting the human on the other side of the screen. Our collaboration has affirmed a core belief of mine: that the next generation isn't afraid of AI, as long as we mentor them to use it with purpose and principle. SCLA aligns with Humanize Innovation in believing innovation should be inclusive, you prove that by reaching students across hundreds of campuses, from all backgrounds, and giving them equal access to opportunity. Thank you for partnering with me to ensure that those who will inherit the AI era are ready to lead it responsibly and brilliantly.

Stedman Graham, Steve Jones, & the Identity Leadership team, your work and your personal guidance have been a cornerstone in my own development as a leader. I still remember the first time I heard you say, "You cannot lead others until you first lead yourself." It was as if a puzzle piece snapped into place for me. Through your Identity Leadership program, you've helped me and countless others define who we are, what we care about, and what we want our legacy to be. This book, in many ways, is part of my answer to the question: "Who am I and what do I have to offer the world?" Stedman, your emphasis on identity and self-awareness has kept Humanize Innovation on course ethically. Whenever we design a new AI workshop or strategy, I can almost hear you asking: "Is this aligned with your vision and values?" That check has never failed us. I want to acknowledge Stedman Graham & Associates for bringing a humanity and clarity to the often-frenetic world of tech innovation. By integrating Identity Leadership into our programs, we haven't just taught clients about AI, we've challenged them to discover themselves, to build confidence and character. That has made all the difference. Thank you, Stedman, for pouring your wisdom into this project and for walking the talk that character is the true bottom line.

I also want to acknowledge the incredible Humanize Innovation team and extended community, the researchers, writers, collaborators, and friends who contributed ideas and time to bring this manuscript to life. To my co-writers and editors: your fingerprints are on every chapter, whether in a sharp insight, a well-timed question, or a meticulously researched fact. You ensured that our work stands on a strong foundation of knowledge and credibility (as evidenced by the extensive sources we cite!). I deeply appreciate your dedication to accuracy and clarity, and for indulging my passion to get the "voice" just right, not too academic, not too hypey, but authentically Greg. Special thanks to those who helped compile the case studies and reference materials in the Sources section; you've created a treasure trove that adds weight to our message.

Lastly, I want to acknowledge you, the reader. By picking up this book (or perhaps reading it on a screen via an AI assistant!), you've become part of the Humanize Innovation journey. Your willingness to explore how we can shape technology to serve humanity gives me hope. Every question you ask, every conversation you start about these topics, creates ripples that make responsible AI and human-centered innovation more of a reality. Thank you for your time and trust. I wrote this for you, and I sincerely hope it equips and inspires you to lead in your own sphere.

In closing these acknowledgments, my voice is one of gratitude and resolve. I am grateful for each partner, colleague, friend, and family member who has contributed to this vision, and I am resolved to carry our shared mission forward. Together, we have built not just a book, but a movement. And this movement doesn't end here on the page; it continues in our workplaces, schools, and communities where we put these ideas into practice. From the bottom of my heart, thank you all for joining me in humanizing innovation.

A Call for Partners, Join the Humaniverse

As we conclude, I extend an open invitation to all readers, organizations, and innovators who see themselves in this vision: Join us as partners in creating a more human-centered future for AI. Humanize Innovation is not a closed club or a finished project, it's a growing community of practice. We call this community the "Humaniverse." The Humaniverse is a collaborative network dedicated to continuing the conversation and turning ideas into action. It's where business leaders, educators, students, technologists, and creative thinkers come together to share insights, pilot new approaches, and keep each other accountable to the highest standards of ethics and impact. By becoming part of the Humaniverse, you signal your commitment to innovation that uplifts humanity. You'll have opportunities to collaborate on research, contribute to workshops and "innovation labs," and connect with like-minded peers who can help you on your journey (and whom you can help on theirs).

The challenges and opportunities presented by AI are far too vast for any one of us to tackle alone. It truly will "take a village," a global community, to ensure AI fulfills its promise without compromising our values. The Humaniverse is our answer to that need. It's more than a community; it's a call to action. In the Humaniverse, every member is both a teacher and a learner, because we all bring something unique to the table. Whether you're a Fortune 500 company reimagining your culture or a student coding your first machine learning project, your perspective matters and can help shape the collective knowledge. We envision roundtables on policy and ethics, open innovation challenges to design AI-for-good solutions, and a mentorship program that pairs seasoned experts with young innovators. And yes, we'll practice what we preach by using great digital tools (including AI) to make our collaboration seamless, but always with a human-first design.

So consider this an open invitation and a heartfelt plea: Become a part of the Humaniverse. Lend your voice, your ideas, and your heart to this movement. You can start by reaching out to us through our website or social channels, where we'll be announcing upcoming community events and forums. Together, we can ensure that the narrative of AI's future is written with human hands and human values. The more diverse and passionate our community, the more powerful our impact will be. Let's partner in this grand endeavor to innovate responsibly, ethically, and inclusively. In doing so, we create a legacy that future generations will thank us for, a world where technology and humanity co-evolve in harmony, each elevating the other. Welcome to the Humaniverse, and thank you for helping build it.

Human First. AI Forward.

That's how we roll.

Featured Case Studies & Partner Organizations

- Hallmark Cards, Inc. — Executive Succession and Leadership Development Case, 2021.
- Identity Leadership — Stedman Graham & Associates. identityleadership.com
- Society for Collegiate Leadership & Achievement (SCLA) — thescla.org
- Rapport — "Using AI to Build Emotional Intelligence in Organizations." rapport.co
- SomethingNew, LLC — "People Over Everything: Award-Winning Talent Strategy." trysomethingnewnow.com
- The Outlier Project (Scott MacGregor) — "Extraordinary Leaders, Extraordinary Impact." theoutlierproject.com
- Heartbeat for Hire (Lyndsay Dowd) — "Top-Down Culture: Building Irresistible Teams." heartbeatforhire.com
- evyAI (Joe Apfelbaum) — "AI-Powered LinkedIn Engagement for Authentic Connection." evyai.com
- Novelle & Associates — "Strategy, Culture & Leadership Consulting." novelle-associates.com

Books & Research Reports

- Bersin, Josh. HR Predictions for 2024: Human-Centered Leadership in an AI World. Josh Bersin Academy, 2024.
- Gartner. "Why 85% of AI Projects Fail." Gartner Research, 2023.
- Harvard Business Review. "Why So Many High-Profile Digital Transformations Fail." Harvard Business Publishing, 2022.
- McKinsey & Company. The State of AI in 2023: Generative AI's Breakout Year. McKinsey Global Institute, 2023.
- MIT Sloan Management Review. "Why AI Will Not Provide Sustainable Competitive Advantage." MIT Sloan Management Review, September 2023.
- NTT DATA. "Overcoming Barriers to AI ROI: Why Leadership Matters." NTT DATA Research Report, 2022.
- Rosenfeld, Bob. The Invisible Element: A Practical Guide for the Human Dynamics of Innovation. Innovating Edge, 2011.
- World Economic Forum. "10 Skills You Need to Thrive in the Fourth Industrial Revolution." Future of Jobs Report, 2020.
- World Economic Forum. "Shaping the Future of Work in the Age of AI." WEF Report, 2021.
- Innovating Edge. "Organizations Don't Innovate, People Do." White Paper, 2020.
- Executive Education at Berkeley. "Emotional Intelligence in the Age of AI." University of California, Berkeley, 2023.

Articles & Emerging Insights
- NTT DATA. "Between 70–85% of GenAI Deployment Efforts Are Failing to Meet ROI." Focus Blog, 2024.

- NTT DATA. "The AI Responsibility Crisis: 81% of Leaders Call for Clearer AI Governance." Press Release, February 2025.
- McKinsey & Company (Mayer, Yee, Chui, Roberts). "Superagency in the Workplace: Empowering People to Unlock AI's Full Potential." McKinsey Report, January 2025.
- Hypermode. "AI Adoption Challenges Unpacked: What's Really Stopping Progress?" Blog Article, March 2025.
- Roland Tomlinson. "Digitally Transformed and Still Broken." Medium, 2024.
- Theresa Stairs. "Humans Wanted: Why Soft Skills Are the Hard Currency in an AI World." True Colors International Blog, July 2025.
- Aspen Institute (UpSkill America). "AT&T Invests $1 Billion in Employee Reskilling." Press Release, March 2018.

About Greg Brisco

Greg Brisco is a leadership strategist, human-centered innovation expert, and trusted advisor to organizations navigating the complexity of the AI era. His work sits at the intersection of leadership, culture, and technology, helping executives, boards, investors, and institutions close the widening gap between technological ambition and human readiness.

Greg is the founder of Humanize Innovation, a platform and practice dedicated to ensuring that as artificial intelligence accelerates, leadership evolves with equal intention. His central belief is simple but often overlooked: technology does not transform organizations, people do. AI is a powerful accelerator, but without aligned leadership, adaptive cultures, and thoughtfully designed teams, even the most advanced systems fail to deliver meaningful results.

Over the course of his career, Greg has worked across corporate, academic, nonprofit, and investment environments, partnering with Fortune-level companies, universities, leadership organizations, and innovation ecosystems. Again and again, he has seen the same pattern emerge. Organizations invest heavily in tools and strategies, yet underestimate the human systems required to execute them. When initiatives stall, the issue is rarely the technology. It is misalignment, outdated leadership models, or cultures unprepared for speed and uncertainty.

This insight led Greg to articulate what he calls the Human Delta™, the gap between bold strategic intent and an organization's actual capacity to deliver. His work focuses on helping leaders identify, measure, and close that gap before it becomes a costly failure point.

Greg's approach is grounded, practical, and deeply human. He challenges leaders to move beyond optimization alone and toward humanization, amplifying the uniquely human capabilities that machines cannot replace: judgment, creativity, empathy, trust, and adaptability. He is a strong advocate for designing leadership systems intentionally, rather than leaving culture and collaboration to chance.

Through Humanize Innovation, Greg helps organizations:

- Engineer leadership alignment at every level
- Build high-velocity, balanced teams
- Govern innovation responsibly in the AI era
- Treat people as strategic infrastructure, not operational afterthoughts

His work integrates research-backed frameworks, real-world case studies, and lived leadership experience. Rather than prescribing one-size-fits-all solutions, Greg equips leaders with lenses and language to make better decisions in dynamic environments.

Greg is also the author of Humanize InnovationI, a leadership manifesto for a world of intelligent machines. The book outlines a human-first, AI-forward approach to innovation, offering leaders a new playbook for navigating change without losing their humanity.

Whether speaking to a boardroom, advising senior executives, collaborating with educators, or guiding emerging leaders, Greg's mission remains consistent: to help leaders rise to the responsibility of shaping technology in service of people, not the other way around.

The future will not be defined by who adopts AI first.
It will be defined by who leads it best.

www.ingramcontent.com/pod-product-compliance
Lightning Source LLC
Chambersburg PA
CBHW071227210326
41597CB00016B/1977